# The Law and Practice Relating to Pollution Control in Ireland

There are twelve other titles in this series:

The Law and Practice Relating to Pollution Control in

Belgium and Luxembourg
Denmark
France
The Federal Republic of Germany
Italy
The Netherlands
The United Kingdom

The Law and Practice Relating to Pollution Control in the Member States of the European Communities: A Comparative Survey
Loi et usage concernant le contrôle de la pollution en Belgique et Luxembourg
Loi et usage concernant le contrôle de la pollution en France
Umweltschutzrecht -und verwaltung in der Bundesrepublik Deutschland
Legge in materia di controllo dell' inquinamento e sua applicazione in Italia

Further details are available from
Graham & Trotman Limited
20 Fouberts Place, Regent Street
London W1V 1HH, United Kingdom.

# The Law and Practice Relating to Pollution Control in Ireland

by Yvonne Scannell, BA, LLB

*Assistant Lecturer in Law, University College, Cork.*

for Environmental Resources Limited

Graham & Trotman Limited
for the Commission of the European Communities

Published in 1976 by

Graham & Trotman Limited
20 Fouberts Place, Regent Street
London W1V 1HH, United Kingdom.
for the Commission of the European Communities.
Directorate-General Scientific and Technical Information and Information Management, Luxembourg.



ISBN 0 86010 031 6



Printed by Page Bros (Norwich) Ltd.

Page

## CONTENTS

This set of volumes was first written in the performance of a contract between the Commission of the European Communities and Environmental Resources Ltd., consultants specialising in environmental research, planning and management. The object was to present an outline of each of the member states' pollution control laws in such a form that they could readily be compared. Supplemented by some of the documents referred to in the volumes, with each national volume in the native language, and English, and the comparative survey in English and French, it was submitted in December 1974, as a reference work to be used principally in the task of carrying out the Communities' programme for action on the environment, and where appropriate for the harmonisation of the member states' pollution control laws.

At the outset it was realised that the volumes might be useful for other purposes, and it was always envisaged that there would be a publication for sale. This is that publication, for commercial reasons slightly more limited in its range of languages. The full list of volumes, with the languages in which they are published, and their authors, appears on the jacket of this volume.

Basically there are nine volumes. Each of the first eight deals with the relevant laws of one member state, except that those of Belgium and Luxembourg appear under one cover. The letter of the law is outlined, and, where possible, an indication is given of the extent of enforcement. The aim is to provide a concise but fully referenced summary. Legislation passed but not yet in force is outlined, and any proposed legislation which has been drafted and placed before the appropriate legislative authority is noted. The ninth, the comparative survey, provides a direct comparison of the relevant laws of all of the member states by presenting them in summary form under the headings 'Air', 'Inland Waters', 'Coastal Waters' etc.

To facilitate comparisons, there has been adopted and used throughout the nine volumes a common classification of controls, a common order of contents, and a common system of numbering the sections and their sub-divisions. Thus, for example, in each volume paragraph 2.2.1.a deals with the control of emissions to and from motor vehicles by means of controls over their design and construction. Two variations from this common pattern must be noted. The first is found in the volume on the Federal German Republic in which, because of the federal constitution, the contents are set out in an extended form of the common scheme of arrangement. The second is that sub-section 8.2 appears only in the volumes on Denmark, Ireland and the United Kingdom. The reason for this is that the Commission was already informed, through Euratom, of the controls over nuclear installations in the original six member states.

It must also be noted that where the legislation of a member state is identical in substance to the provisions of an international convention, that legislation is not necessarily summarised; the reader is in some cases left to refer to the convention. This stems from the fact that the work was originally for use by the staff of the Commission, and there was no need to include the details of provisions which were already well known to them.

This present publication is offered with two principal purposes in view. It enables the reader to study in outline the pollution control system of any one member state. It also enables him to make a direct comparison of the controls of any two or more member states, either generally or in some particular field. This comparison can be made either by using the comparative survey, or in more detail by referring to the corresponding paragraphs in the national volumes.

The arrangement of contents thus adapted for the nine volumes may not be the best for examining or explaining the controls of any particular country. For some countries, indeed, because of their different systems of law, the structure of their control authorities or the forms of control of which they principally rely, an entirely different approach would probably have been more suitable. But the main purpose was comparison and a classification of controls appropriate to that purpose had to be devised. Moreover the Commission officials had emphasised their interest in the enforceability of the various types of control.

I had always been convinced that controls could be classified, and the ease with which they could be effectively enforced compared, by dividing into stages the processes by which pollution is generated, and by classifying the controls according to the stages at which they are applied. These stages begin with the planning of any new development, proceed through the various stages of manufacture, discharge of process wastes and use of the product, to the last stage at the final disposal of a consumer product. An arrangement of contents based largely on this classification was prepared, and submitted for discussion at a meeting of the authors. It was accepted after some useful modifications had been made.

I am most grateful to the authors for their willing spirit of cooperation, and for the generous way in which they accepted and adapted their work to a fairly rigid structure. There is little doubt that in every case, sometimes as a result of this structure, the author's account of his own state's laws is not as coherent as it would otherwise have been, but by this compromise we have gained a means of direct comparison. For any defects or deficiencies in the arrangement, I accept full and sole responsibility.

Presenting a nation's laws accurately in summary form is always a difficult task. There is a danger that, out of context, they may be misunderstood – that is why we have attempted to put them in context by giving, in section I, something of the constitutional and legal background. A further danger lies in translation. Although we have tried in each case to render as accurate a translation as possible, we must emphasise that only the version in the native language carries the authority of the author of that volume. The comparative survey is my responsibility alone. It is offered solely for the purpose of comparison; for other purposes recourse should be made to the national volume.

The statement of law in each volume is correct to 30 May 1974. For speed in publication it is offered without any revision or updating but it will be followed by a supplement to each volume making the statement of law correct to 30 October 1975.

May I acknowledge again, and in a more general sense, the ready cooperation of all the various authors. I was told that it would not be easy to get people of so many nationalities to agree to a common form, and that it would be impossible to get them to produce their work within close time constraints. Both those prophesies were proved wrong. It now seems surprising that we finished the work project, including translation and presentation of the report to the Commission within eighteen months.

Finally, I should like to acknowledge the assistance of many agencics, governmental and non-governmental, which have freely given information and advice, and the help and guidance given by Mrs. M. Brusasco and other members of the staff of the Environment and Consumer Protection Service, The Commission of European Communities.

J. McLoughlin LLM
Associate, Environmental Resources Limited

# Principal Pollution Control Legislation in Ireland

## Pollution in General

Local Government (Planning and Development) Act, 1963

### 1. Air Pollution

Alkali etc. Works Regulation Act, 1906.
Local Government (Sanitary Services) Act, 1962.
Control of Atmospheric Pollution Regulations, 1970.
Forehsore Act, 1933.
Minerals Development Act, 1940.
Petroleum and Other Minerals Development Act, 1960.
Road Traffic Act, 1968.
Road Traffic (Construction Equipment and Use of Vehicles) Regulations, 1963 as amended

### 2. Water Pollution

Fisheries (Consolidation) Act, 1959.
Fisheries (Amendment) Act, 1962.
Rivers Pollution Prevention Act, 1876.
Public Health (Ireland ) Act, 1878.
Public Health Acts Amendment Act, 1890.
Harbours Acts 1946, 1947.
Foreshore Act, 1933.
Oil Pollution of the Sea Acts, 1956, 1965.
Continental Shelf Act, 1968.

### 3. Wastes on Land

Local Government (Planning and Development) Act, 1963.
Public Health (Ir.) Act, 1878.
Road Traffic Act 1968.
Minerals Development Acts 1940, 1960.

### 4. Noise

Public Health Acts Amendment Act, 1890.
Air Navigation and Transport Act, 1950.
Factories Act, 1965.
Factories (Noise) Regulations, 1975.

### 5. Nuclear Energy

Nuclear Energy Act, 1971.

### 6. Products

Fertilisers, Feeding Stuffs and Mineral Mixtures Act, 1955.
Diseases of Animals Act, 1966.
Poisons Act, 1961, (Paraquat) Regulations, 1975.
European Communities (Detergents) (No.2) Regulations, 1975.

## Table of cases

*Dunne Ltd.* v *Dublin County Council,* (1974) IIR.45, at p.53.
*Crossley* v *Lightower,* LR 2 Ch App. 478.
*Killiney and Ballybrack Development Association Ltd* v *Minister for Local Government and Templefin Estates Ltd.,* HC 1.3.74 (unreported)
*Mullin* v *Hayes,* SC 13.11.72
*Callin* v *McAvinue,* HV 11.5.73 (unreported).
*McDonald* v *Associated Fuels* (1954) 3 DLR.775.
*Ryan* v *Attorney-General* (1965) IR at p.344.

# Introduction

*We have attempted to give a comprehensive description of all types of pollution controls __ legal and extra - legal within the framework of the terms of reference. This task was more difficult than may be apparent because of the dearth of literature on the subject in this country. It will be seen that such controls as do exist closely resemble British controls in the same field and that the Local Government (Planning and Development) Act, 1963, is in practice the main instrument of pollution control.*

*Yvonne Scannell*

# 1 The nation's authorities and pollution control

1.1 Overall national government in Ireland lies with the *Oireachtas* which, according to the constitution, consists of the President and two Houses, viz. a House of Representatives — the *Dail* — and a Senate. The *Oireachtas* has the 'sole and exclusive powers of making laws for the State'. Its legislative power is limited by the provision of the constitution and by its own organisation and procedures. It has been said that this means that 'in comparison with other democratic countries, its role in government is modest' because a government backed by a majority in *Dail Eireann* is able to manage the conduct of its business to suit itself and has a virtual monopoly of initiative in proposing bills and other motions[1]. The constitution provides that 'the executive power of the state shall be exercised by or on behalf of the government.' The *Oireachtas* lays down by legislation the broad lines of policy to be carried out, and finances the execution of these policies. It also has the constitutional power to review the execution of its policies, and executive action may be questioned and criticised by means of parliamentary questions and by debates on the annual estimates. In practice, the ability, effectiveness and, indeed, willingness of the *Oireachtas* to control, supervise or enquire into the administration of the country is severely limited by out-dated methods and procedures, by the lack of a developed committee system as exists in Britain, and by the appalling facilities available for the use of any parliamentarian who would venture into the administrative domain.

Justice is administered in courts established by law. The District Court has a criminal jurisdiction embracing

(i) minor offences triable in a summary way,
(ii) indictable criminal offences specified in appropriate legislation which are triable summarily with the consent of the accused,

**1.1**
**continued**

(iii) indictable offences (other than treason, murder or piracy) where the accused pleads before the District Court and where the Justice is satisfied that he understands the charge,

(iv) indictable offences not triable summarily where there is a preliminary examination by the Justice,

(v) certain matters of a quasi-criminal nature conferred on the District Court by statute e.g. granting dance-hall licences or licences under the Intoxicating Liquor Acts.

The Circuit Court has the same criminal jurisdiction as the Central Criminal Court save for treason and analogous offences, murder and piracy.

The Central Criminal Court is the High Court exercising its criminal jurisdiction, and it has original criminal jurisdiction to try crime on indictment.

Appellate criminal jurisdiction is enjoyed by the Circuit Court which hears District Court appeals; the High Court, which may issue orders to inferior tribunals for the purpose of bringing up their orders for review, and which may give rulings on questions of law submitted to it by a District Court; the Court of Criminal Appeal which hears appeals from the Central Criminal Court and Circuit Courts, and the Supreme Court which may

(i) hear appeals from the Court of Criminal Appeal,

(ii) may decide on points of law certified by the Attorney General or by the Court of Criminal Appeal as involving points of law of 'exceptional public importance,'

(iii) may hear appeals against High Court prerogative orders or the refusal of same,

**1.1** **continued** (iv) may hear appeals against High Court decisions with respect to cases stated from the District Courts, and

(v) may entertain cases stated from Circuit Courts.

Civil jurisdiction is exerciseable by the District Court in contract and tort where the claim does not exceed £250, and in proceedings at the suit of the state, or any official thereof, where the amount involved does not exceed £250.

The Circuit Court has civil jurisdiction in contract and tort where the amount claimed does not exceed £2000, the High Court has unlimited original jurisdiction in all civil matters, and on questions as to the validity of any law having regard to the provisions of the constitution.

Appeal from a District Court decision lies to the Circuit Court. There may be an appeal to the High Court by way of 'case stated' on a point of law, and a further appeal, by leave of the High Court, to the Supreme Court. Appeal from Circuit Court lies to the High Court, and the High Court when hearing such appeal may state a case for the Supreme Court on any question of law. The Supreme Court is an appellate court High Court decisions or orders, and it hears 'cases stated' by Circuit Courts.

It is unfortunate that the vast majority of cases which involve pollution are heard in the District Court, the decisions of which are not usually reported. The suitability of this court as a tribunal for hearing matters relating to pollution and planning is extremely questionable.

The administration of the country is primarily the function of the sixteen government departments, each headed by a minister who is personally and constitutionally responsible for all the activities of his department. These departments advise the Government on policy matters,

**1.1**
**continued**

review the operation of and carry out the detailed execution of existing policies. Below this level there are three principal types of bodies with administrative functions — local authorities, state-sponsored bodies, and advisory and other committees.

**Local authorities**[2]

The strict legal and constitutional position in Ireland is that local authorities are bodies charged by the *Oireachtas* with the responsibility for the performance of certain duties in particular areas. They derive their power, functions and dutes from legislation, and are themselves creatures of the law with only such existence, powers and functions as the law confers.[3] The Public Services Organisation Group has commented that 'a feature of the local government system is the high degree of central control and direction applied to it.'[4] The ultimate sanction lies in fiscal measures since more than half the finance of local authorities comes from the National Exchequer.[5]

**State-sponsored bodies**

State-sponsored bodies in Ireland vary enormously in their structure, size, and in the roles which they are called on to play. Their establishment was largely 'a consequence of the extension of government activity into fields which were previously the field of private commercial enterprise.'[6] There are two basic types of state-sponsored bodies, viz: those which exercise a commercial function on behalf of the state, and those which are purely executive agencies of the government. The overall picture is of a 'central core of departments with their association fringe of agencies reporting to those departments with which they are functionally connected.'[7] Some[8] are statutory corporations, others[9] are public or private companies, or occasionally companies limited by guarantee. Differences in practice between bodies are more apparent than real, and result more from the personalities involved in them, than from their various organisational forms.

**1.1 continued** Central government control operates at three levels — financial, parliamentary and when appointing the boards. The degree of financial control varies with the body involved, because some rely entirely on the state for their finance, others are partly subsidised by the state while many of the trading enterprises are self-supporting and obtain no direct subsidy from the state. In theory, state-sponsored bodies are responsible to their ministers, and through their ministers to the *Oireachtas*. Their annual reports and accounts are presented to the *Oireachtas* and may be subjected to criticism and debates by resolution. In addition, periodic parliamentary questions may enquire into the activities of a body. By convention, however, the day-to-day activities of these bodies is their own business. The boards of most state-sponsored bodies are appointed by the government minister or ministers who are responsible to the *Oireachtas* for their operations. The main role of the boards is to implement government policy and to maintain liaision with their particular minister. The relationship between boards and ministers is not always clearly defined and 'a great deal depends on the confidence which a particular minister may have in the board or executives of a particular concern and on the readiness of a board to accept that the particular operations for which they are responsible should be kept in line, through the minister as the coordinating authority, with the overall development plans of the Government.'[10] The minister's main instruments for ensuring compliance with his policy are criticism, informal communications and, as a last resort, dismissal of an unsatisfactory board.[11]

The role of state-sponsored bodies in the field of pollution control will be discussed in 1-5.

### Advisory and other committees

The committees are as numerous as they are varied. They consist of members, largely appointed by the ministers for

1.1 continued

the purpose of investigating local or national situations, or particular subjects, and advising them thereon. Such committees may be established under a power given in a statute or as a result of the personal initiative of a minister. They have no legislative or administrative powers but they may, and do, exercise persuasion at other levels.

The law relating to pollution derives from three principal sources:

1. the constitution,
2. the common law, and
3. statute law.

The constitution is the source from which all laws derive their validity. In addition, it guarantees certain fundamental rights, which may not be infringed. The common law is the source of certain criminal offences, and it seeks to protect and enforce the legitimate interests of individuals. Statute law together with subordinate legislation made thereunder (e.g. regulations and bye-laws) is a major source of the law relating to pollution.

## 1.2.1 The Constitution

Article 40.3.2 of the constitution provides that 'the state shall in particular, by its laws, protect as best it can from unjust attack and, in the case of injustice done, vindicate the life, person, good name and property rights of every citizen.' It has been held[12] that the personal rights mentioned in the aforesaid subsection are not exhausted by the enumeration thereof, and that one of the unmentioned personal rights is the right to 'bodily integrity.' This right has been judicially defined to mean 'that no mutilation of the body or any of its members may be carried out on any citizen under authority of law except for the good of the whole body and that no process which is or may, as a matter of probability, be dangerous or harmful to the life or health of the citizens or any of them may be imposed (in the sense of being made compulsory) by an Act of the *Oireachtas.*'[13] A cogent argument could be advanced, on the basis of the *Ryan* v. *A.G.* case, that there exists in Ireland, a right not to be polluted.

## 1.2.2 Common law

The only common law criminal offence which can be committed merely by polluting the environment is the offence of the public nuisance. Public nuisance is an act which interferes with the enjoyment of a right to which all members of the community are entitled. It may also be a civil wrong. In theory, it could cover many forms of pollution. In practice it is rarely used. The remedies granted to a successful plaintiff are normally damages and/or an injunction.

## 1.2.3 Statute law

Legislation enacted for the purpose of preventing pollution usually seeks to achieve this by creating offences punishable by fines and/or imprisonment. Sometimes a statute seeks to reinforce the common law, other times it seeks to create new duties, i.e. 'statutory duties.' Whether or not breach of a statutory duty gives rise to a right of action at the suit of an individual depends upon the construction of the particular statute. Very few statutes were passed to deal specifically or exclusively with pollution, but public health legislation provides for the punishment of a number of 'statutory nuisances' which embrace various types of pollution, and the power to abate them is given to local authorities where such action appears to be necessary in order to protect the health of the community. Such 'statutory nuisances' need not necessarily interfere with personal comfort but they can only arise from a private source and they cannot (by definition) arise from public works, such as sewage works.

**1.2.3 continued** The following are examples of statutory nuisances created by the public health and other legislation.

1. Any premises in such a state as to be a nuisance or injurious to health.

2. Any pool, ditch, gutter, watercourse, privy urinal, cesspool, drain or ashpit so foul or in such a state as to be a nuisance or injurious to health.

3. Any animal so kept as to be a nuisance or injurious to health.

4. Any accumulation or deposit which is a nuisance or injurious to health.[14]

5. The drainage system of one or more premises which is so defective, foul or neglected as to be or to be likely to become a nuisance or injurious to health.[15]

6. Temporary dwellings in such a state as to be a nuisance or injurious to health.[16]

7. Pollution of the air by premises other than private premises.[17]

All these and other statutory nuisances may be prosecuted by sanitary authorities and every inspector or other member of the police is required to report all such nuisances as shall occur within his district to the district justice at his next sitting in the district,who shall punish such nuisance by fine or otherwise as may be directed by law.[18]

While it is clear that statutory nuisances do not extend to everything that could be a nuisance at common law the words 'nuisance or injurious to health' are not confined to nuisances affecting health, but include such things as diminish the comfort of life though they may not be injurious to health, the word 'or' being used disjunctively.[19]

Conviction for statutory nuisance is a means of punishing pollution but the penalty provided is usually too minimal to constitute an effective deterrent.

## 1.3 GOVERNMENT DEPARTMENTS

No single government department or agency has overall responsibility for pollution control in Ireland. This task is shared jointly or severally by a number of government departments. Of these the Department of Local Government is by far the most important and it approximates most closely to what might be called a Department of the Environment. The necessary coordination and cooperation between departments is secured through discussion, consultation, liaison-committees and exchanges of information on a formal or informal basis — largely the latter. A notable example of the former was the setting up in 1970 of an inter-departmental working party, comprising representatives of seven departments, to report to the Minister for Local Government on problems relating to air and water pollution.[20]

The diffusion of responsibility for pollution controls is confusing and sometimes illogical. The following paragraphs indicate the principal aspects of pollution control which are the responsibilities of the various departments:

### (i) Local government

General responsibility for air and water pollution, oil pollution of coastal areas, the preparation and implementation of development plans, the control of development, the preservation and improvement of amenities, and local traffic. The department is also the central control authority

1.3 continued for local authorities in their various capacities and for *An Foras Forbartha.* The department has special sections dealing specifically with planning, environmental services, and sanitary services.

(ii) **Transport and power**

Responsibility for transport, energy, oil pollution of the sea, tourism, foreshores, meteorological services, harbour authorities, the Electricity Supply Board, *Bord Failte.*

(iii) **Agriculture and fisheries**

Responsibility for silage effluent, farmyard waste disposal, artificial fertilisers and pesticides, creameries, some food processing factories, protection of fish and food forms in water, and of spawning grounds and fish as food, licensing discharges deleterious to fish life. The department is the central control authority for boards of fishery conservators, for the Inland Fisheries Trust, the Salmon Research Trust of Ireland and An Foras Taluntais.

(iv) **Industry and commerce**

Granting prospecting licences, permissions and leases in respect of state-owned minerals, licences for the manufacture of cement, and smelting licences, responsibility for industrial development, control and supervision of the Institute of Industrial Research and Standards, and the Industrial Development Authority.

(v) **Health**

Concern with environmental health problems, and food contamination, overall supervision of health boards.

(vi) **Lands**

This department is concerned with the pollution effects on wild-life, and water pollution aspects of forestry development.

The price paid for this overlapping of functions in relation to pollution control is diffusion of responsibility, inefficiency, departmental rivalry, the lack of clearly defined roles and responsibilities and the inhibition of progressive developments. There have been repeated calls for a separate Department of the Environment, but the Government does not intend setting up such a department.

## 1.4 Regional and local pollution control authorities

The regional and local pollution control authorities of various kinds belong mostly to the local government system, or are more or less closely allied to it. They include:

1. Local authorities
2. Planning authorities
3. Regional development organisations
4. Sanitary authorities
5. County committees of agriculture
6. Boards of fishery conservators and
7. Harbour authorities

All these bodies have a territorially limited jurisdiction and function for regions or counties or sub-county districts, or in the case of harbour authorities, for defined port areas. Together they constitute a loosely articulated system of local administration.

### 1.4.1 Local authorities

The present system of local government has its origins in nineteenth century legislation. The elected authorities are county councils (twenty-seven), county borough corporations (four), borough corporations (seven), urban district councils (forty-nine), and boards of town commissioners (twenty-eight).

### 1.4.1.a Constitution

County councils and county borough corporations were established by the Local Government (Ireland) Act 1898. Borough corporations were reformed and re-constituted by the Municipal Corporations (Ireland) Act 1840. Town commissioners came into existence in some cases under an Act of 1828, in others by Private Act, but generally under the Towns Improvement (Ireland) Act, 1854. The modern urban district council is the successor of the urban sanitary authorities whch were created by the Public Health (Ireland) Act, 1878.[21]

Members of these authorities are elected, normally every five years, on a system of proportional representation. Elections are fought almost exclusively on a party political basis. All qualified persons registered as local government electors are entitled to vote. Elected members have overall responsibility over the activities of their particular local authority, and certain exclusive powers, while administrative matters are left to a manger and staff. Membership of these bodies fluctuates; county councils may have twenty to forty members, county borough corporations fifteen to forty-five, borough corporations twelve, urban district councils and town commissioners nine.

**1.4.1.b** County councils exercise their jurisdiction over that part of a county unit which is not within the jurisdiction of other local authorities.[22] County borough corporations are responsible for the cities of Dublin, Cork, Waterford and Limerick. Borough corporations, urban district councils and town commissioners have functions over a number of smaller towns.[23]

**1.4.1.c** The White Paper on Local Government[24] describes the powers of local authorities as being 'mainly concerned with physical planning and development, the provision of

1.4.1.c continued infrastructure, environmental services, amenity services, and to a lesser extent, with protective and regulatory services.'[25]

County councils, county borough corporations, borough corporations and urban district councils are *planning authorities* under the Local Government (Planning and Development) Act 1963, and *sanitary authorities* under the Local Government (Sanitary Services) Act 1962.[26] They are empowered to make bye-laws for the good rule and governments of their jurisdictions and for the prevention and suppression of nuisances.

**1.4.1.d** Local authorities have a very close relationship with central government and especially with the Department of Local Government. This relationship has been described as 'a special form of partnership,'[27] but if this is true, the Department is very much the senior partner. The Minister for Local Government is responsible to the *Dail* for the activities of local authorities and central government control is extensive. The Minister has general tutelary powers over certain services, is responsible for applying the law by rules, orders and regulations, maintains special advisory organisations, has many powers including a power of approval in relation to many local authority plans, a power of action in default and of invoking the control of the law courts, to approve all local loans, to make enquiries, and obtain reports, etc.[28] He has also the extraordinary power of removing members of local authorities from office in specially legally-defined circumstances. Most important of all, perhaps, local authorities get more than fifty per cent of their finances from central funds.

Local authorities may, and frequently do, cooperate with or seek the assistance of each other, and of other statutory bodies.They may exercise or perform a function, power or duty of another local authority or statutory body,[29] may establish corporations to perform services for them,[30] and may appoint as such and as many advisory committees as they think fit, to consider any matters connected with local authority functions.

1.4.1.e The effectiveness of local authorities in the task of controlling pollution is extremely doubtful. It has been alleged that sixty-five per cent of water pollution in Ireland is caused by local authorities and many local authority dumps contribute to the pollution problem. It is ironic therefore that these bodies be charged with such responsibility for pollution control, and that their own development is not, as a rule, subject to the provisions of the Local Government (Planning and Development) Act, 1963. They are often over-anxious to attract industries to their jurisdictions and are apt to be over-liberal in the latitude, in environmental terms, which they allow such industries. Political rivalries, and conflicting interests when dealing with pollution control matters do not contribute to effectiveness of their role in protecting the environment. They lack financial and technical resources, manpower, and the will to undertake the role of pollution controllers.[31] They are themselves among the worst polluters.[32] The rate of progress between different authorities varies, and much depends upon the finance, philosophy, iniative and forcefulness of each particular authority.

A complete re-organisation of local government is being considered.[33]

## 1.4.2 Planning authorities

Under the Local Government (Planning and Development) Act, 1863, county councils, county boroughs borough corporations and urban district councils — eighty-seven bodies in all — are planning authorities for their respective districts.

Planning authorities have two basic fields of activity:

**1.4.2.a** Planning by means of restrictive controls as to new structures, or as to the alteration of structure or use of existing buildings or land, exercised on an ad hoc basis; and

**1.4.2.b** Planning in positive form; that is, by creation of 'development plans' whereby the future use of land in every planning district may be chartered or ascertained.

**1.4.2.c** The Act requires that each local planning authority prepare a plan indicating the development objectives for their areas. These plans are policy statements showing the ways in which land is to be apportioned between competing claims for housing, commercial and other uses, so that development is facilitated and conflicting uses kept reasonably apart. The plans for cities and towns are to show land use zones, proposals for dealing with future traffic, for development and renewal of obsolete areas and amenity areas for preservation. Plans for counties must show objectives for the development and improvement of the county including the preservation of amenities, the redevelopment of obsolete areas, the provision and extension of water and sewage services and the improvement of traffic conditions in the towns. Planning authorities are expressly empowered to indicate a wide variety of other objectives, including objectives for pollution control.[34] Development plans must be renewed at least every five years and are available to the public.[35] Copies of the plans must be submitted to the Minister for Local Government, who may require that they be varied, and/or that the development plans of two or more planning authorities be coordinated in the mode and in respect of matters specified by him.[36] In addition, he may publish general instructions in relation to the preparation of such plans together with model forms of development plans and their contents.[37] Such plans must be publicly displayed for at least three months and the planning authority is bound to consider objections and representations received. Planning authorities are obliged to take such steps necessary for

1.4.2.c continued securing the objectives of development plans[38] and the plans operate as a framework within which applications are made and approvals given.

Planning permission is required for all development, not being exempted development or development commenced before the appointed day, and for the retention of unauthorised structures which existed before that day.[39]

Development is defined as 'the carrying out of any works on, in, or under land, or the making of any material change in the use of any structures or other land.'[40] Exempted development is extraordinarily widely defined, in the Act itself, and in subsequent regulations and includes, *inter alia,* almost all development on agricultural land and on land used for forestry, and most development by local authorities.[41] The Minister for Local Government is empowered to, and has made regulations providing for the grant of permissions for the development of land and for the retention of the aforementioned unauthorised structures.[42] These regulations provide for the grant of permissions for the development of land and the retention of structures which are unauthorised structures on the appointed day. They also provide for the giving of outline permissions for development subject to the subsequent approval of the planning authority, for the giving of notices in relation to applications for permission, and for plans and information to be given to the planning authority by applicants for permission. All applicants must give public notice of intention to apply for planning permission or approval by publication in a newspaper circulating in the district in which the land or structure is situated, or by fixing a notice on the land or structures.[43]

Planning authorities may decide to grant *permission* for development or for the retention of unauthorised structures 'subject to or without conditions', or to refuse such permission.[44] While the discretion of the authorities as

1.4.2.c continued to the conditions they may couple with planning permissions is not entirely absolute;[45] it is taken to be wide enough to permit the imposition of conditions to prevent, or minimise, all types of pollution.

Planning authorities may also grant *licences* to erect, construct, place and maintain certain appliances or structures on, under, over or along a public road, for such period and upon 'such conditions as they may specify', and may, in defined circumstances withdraw such licence by written notice and require the licensee to remove such appliance or structure at his own expense.[46] In certain circumstances planning authorities may grant *permission* to dispense with, or relax any requirement in building regulations.[47] A grant of permission unless it provides to the contrary, ensures for the benefit of the land or structure and of all persons interested in it.[48] It is effective only for the purposes of the Planning Act and does not relieve the developer of the necessity of complying with other legislation affecting the use or development of land.

*Conditions attached to planning permissions are potentially, if not in practice, the principal form of pollution control in this country.*

Development carried out in breach of the planning code is an offence, punishable on summary conviction by a fine not exceeding £20, and if the offence continues after conviction, by a fine not exceeding £10 for each day during which the contravention is continued.[49] Provision is also made in the Act for the punishment of other offences.[50] Prosecutions may be brought by the planning authorities in whose area the offence is committed.[51]

Planning permission may be revoked or modified by notice, in certain circumstances, if the planning authority decides it is expedient, having regard to the proper planning and development of their jurisdictions and other

**1.4.2.c continued** material considerations, but such revocation or modification may not affect work which has already been carried out, and the authorities may have to purchase the property concerned.[52]

Planning authorities have power to serve enforcement notices for breach of the planning code.[53] The law relating to the service of such notices is highly technical. When development is carried out without permission, or where a condition or limitation attached to a planning permission has not been complied with, a planning authority may, on its own initiative, and must if directed by the minister, serve an enforcement notice on the owner or occupier of the land. Such notice must be served within five years of the breach of the planning control otherwise the breach becomes immune from action. Before serving the enforcement notice, the authority must be satisfied that it is expedient to serve such notice, having regard to the proper planning and development of its area, and other material considerations. The notice must specify the alleged breach of the planning control and may specify the period within which the breach is to be remedied and the steps to be taken to remedy the situation. Such steps may require, *inter alia,* the removal of any structure. Non-compliance with the terms of the notice entitles the authority to enter on the property concerned and carry out the requisite works themselves at the expense of the owner of the land. Particulars of all such notices must appear in the register kept by the authority.[54] Almost identical provisions apply for breach of any condition subject to which a permission for retention of a structure has been granted.[55] Where any development authorised by a permission has been commenced and has not been, or is not being carried out in conformity with such permission the planning authority may serve an enforcement notice on the person who commenced the development, or on any other person who has carried out or is carrying out development authorised by the permission. There is no time specified within which this enforcement notice must be served.[56]

**1.4.2.c continued** If the planning authority decides, having regard to material considerations, that any structure should be removed or altered, it may serve a *notice* on the owner or occupier of the structure requiring such removal or alteration, and in the case of a removal, requiring any replacement that it may consider suitable. Non-compliance with the requirements of such notice entitles the planning authority to enter on the structure concerned and carry out the requisite works.[57] The authority may have to purchase the property concerned.[58] Somewhat similar provisions apply where a planning authority decides as regards any use of land that the use be discontinued or that any conditions should be imposed on the continuance thereof, except that the planning authority may not take the required action itself.[59] Unfortunately 'use' in relation to land does not include the use of the land by the carrying out of any works thereon.[60]

Refusal to allow a planning authority to exercise a power of entry conferred by the Act obliges the authority to obtain a District Court order approving such entry. Obstruction of entry approved by such order is an offence.[61]

A planning authority may decide that it is expedient in the interests of an amenity that any hedge may be removed or altered and may serve a notice requiring such removal or alteration or suitable replacement for a removed hedge. If the removal or alteration is not affected as required, the planning authority may take the required action itself.[62]

In addition to the general power of control, planning legislation provides a number of special forms of control concerned with preserving or improving the environment, e.g. planning authorities may, as permitted by the Act, make amenity orders,[63] tree preservation orders,[64] conservation orders,[65] may plant or assist those who would plant trees, shrubs or other plants,[66] and may serve notices requiring persons to repair or tidy advertisement structures or advertisement notices over which they have control.[67] They may also develop or secure the development of land for a wide variety of purposes and may provide sites for industries and other services.[68]

1.4.2.d Planning authorities work closely with the Minister and Department of Local Government and are in regular receipt of recommendations and advice from these bodies. They also work, when the situation warrants it, with other departments concerned with pollution control. They cooperate with each other[69] and with regional development and county development teams. They constitute a part of every local authority and work extremely closely with other local government authorities, especially the sanitary services authorities. Indeed the latter are often housed in the same buildings as the planning authorities. A certain amount of communication exists between planning authorities and state-sponsored bodies, especially the Industrial Development Authority, the Institute of Industrial Research and Standards, *An Foras Forbatha* and *An Foras Taluntais,* all of which provide technical advice to planning authorities.[70] Planning authorities may, and occasionally do, assist local development associations; companies concerned with the tourist industry in tourist resorts; and any person or body concerned with the preservation or development of amenities, carrying out works of local improvement or maintenance of amenities, and sometimes, town commissioners. Such assistance may consist of helping in money or in kind or of providing services and facilities.[71] The Act obliges planning authorities to consult certain 'prescribed authorities' in stated circumstances.[72] In practice, copies of applications for planning permission which are relevant to pollution control are sent to interested bodies.

The effectiveness of the Planning Act as a measure to control pollution is not complete. There are many loopholes

**1.4.2.d continued** in the Act because many sections were copied from British legislation without the backing up and ancillary pieces of legislation in Britain which form a part of the total code there. It has been alleged that 'the Planning Act in its present form discriminates in favour of classes, individuals and statutory bodies with wealth and influence against the interests of the community as a whole.'[73] Indeed, there are very significant exemptions from development which can give rise to disquiet,[74] and there are compelling reasons for controlling much exempted development and in particular agricultural development. The Act grants exemption in retrospect to certain developments on the basis of laws which it was designed to replace, thus making it possible in certain circumstances to inherit the right to dispensation.[75] Sufficient time has now elapsed for the transitional period to be deleted and for all permissions granted under the old Acts to be declared null and void. Conditions attached to planning permissions are not subject to review in the light of changing circumstances which may require that qualitative and quantiative standards be varied. Permissions should not automatically be granted as a result of default on the part of the planning authority since delays in notification can be caused by reasons other than indolence, e.g. strikes.[76] Experience with the operation of the enforcement provisions of the Act shows that they are too cumbersome and complicated to deal quickly and effectively with breaches of the Planning Code. In any case, local authorities lack the personnel to police the operation of permissions and are reluctant to use their control powers by prospects of paying compensation,[77] or of being served with purchase notices.[78] The legal validity of some conditions imposed to control pollution is questionable, and express provision should be made for the imposition of such conditions. Planning authorities are frequently subjected to political pressures advocating the imposition of lenient conditions in the interests of attracting industries and providing employment. One of the major shortcomings of the Act is that it provides virtually no provisions for

**1.4.2.d continued** controlling pollution by development carried out or commenced before 1 October 1964.

Enforcement of the Planning Code is haphazard and infrequent. The following table, adapted from the Report on Water Pollution[79] indicates that the potential of the Planning Act to control pollution is not being exploited. It consists of a summary of information from a survey carried out by the Department of Local Government of planning permissions granted for industries liable to give use to effluent problems.

| | |
|---|---|
| No. of planning authorities | 87 |
| No. of replies to date | 42 |
| No. of processes approved | 60 |

**Conditions governing effluent imposed**

| | |
|---|---|
| Yes | 34 |
| No | 9 |
| No information | 17 |

**System of checks in operation**

| | |
|---|---|
| Not applicable | 9 |
| Periodic testing or inspection | 8 |
| Monitoring of effluent | 6 |
| No checks | 2 |
| Inspection of completed works | 1 |
| No information | 34 |

**Effluent treated**

| | |
|---|---|
| Yes | 19 |
| No | 21 |
| No information | 20 |

**1.4.2.d**
**continued**

**Complaints of pollution received**

| | |
|---|---|
| Yes | 6 |
| No | 22 |
| No information | 32 |

**Proposed new legislation**

The Local Government (Planning and Development) Bill, 1973, if enacted will go some way towards improving the effectiveness of the Planning Act. The main purposes of the Bill are:

(i) to provide for the establishment of a new board — to be known as *An Bord Pleandala* — to deal with most of the appeals references, and other matters, now dealt with by the Minister for Local Government, under the Local Government (Planning and Development) Act, 1963. While members of the proposed new board will be appointed by the Government and by the Minister for Local Government, the Bill tries to ensure that it will act impartially, and that its members and employees will be free from political and other influences;

(ii) to strengthen some provisions of the 1963 Act. The Bill, *inter alia,* gives planning authorities greater powers to take action to secure the proper completion of housing estates where open spaces have not been provided as indicated in the developer's application for permission or as required by conditions attached to planning permissions, and provides a new enforcement procedure to discourage unauthorised development. In particular, the Bill provides that a planning authority or any person may apply to the High Court, seeking an order to prohibit the continuance of any unathorised development or use of land;

(iii) to make miscellaneous amendments to the original Act, including changes in the law relating to development plans, noise, building regulations, compensation, procedure, conservation orders, and requiring that planning permission be sought for the fencing or enclosure of land which has long been open to, or used by, the public.

## 1.4.3 Regional development organisations

Regional development organisations are non-statutory bodies representative of local authorities and of other development interests such as the industrial and tourist authorities in the nine planning regions. Membership includes both elected members and officers of local authorities. The organisations report on, *inter alia,* physical and infrastructural factors which are important for regional development programmes. Their main interest is economic development, but a number of organisations have commissioned reports on pollution problems and matters ancillary thereto,[80] and have drawn up plans to deal with oil pollution on a regional basis. Reports of regional development organisations are taken into consideration by planning and other pollution control authorities, especially when locations suitable for industries are being examined.

## 1.4.4 Sanitary authorities

County councils, county borough and borough corporations and urban district councils are sanitary authorities for their areas. Their powers, functions and duties are governed by a vast number of Acts dating from the early nineteenth century. They are concerned with the provision of water supplies, sewerage and drainage, refuse collection, waste disposal, street cleansing and the elimination of conditions prejudicial to a healthy environment. They have the necessary powers to carry out these functions, and local authorities may, acting as sanitary authority, make laws for controlling public nuisances.[81]

Sanitary authorities themselves are major polluters, largely because of the inefficient and antiquated methods

**1.4.4 continued** employed for disposing of sewage. Virtually all sewage systems serving populations of more than 10,000 are causing or contributing to pollution and require remedial action.[82] They have power to prosecute for certain statutory nuisances[83] and are charged with enforcing the Control of Atmospheric Pollution Regulations, 1970.[84]

Sanitary authorities cannot be expected to fulfil their obligations in the proper manner unless they are provided with the finance to do so. The cost of carrying out remedial and improvement works on sewage systems causing pollution was estimated at £35 million in 1972.[85] The total capital provision given for all sanitary services' programmes in 1971-1972 was £6.25 million.[86]

Sanitary authorities work closely with the Minister and Department of Local Government, with planning and housing authorities and with all persons and bodies for whom services are provided.

## 1.4.5 County committees of agriculture[87]

Under the Agriculture Act, 1931, as amended, each county council must appoint a committee of agriculture. Each committee has corporate status. Members are appointed by the county council and may include persons who are not members of the council. The committee appoints its own officers and normally has a chief agricultural officer as its chief executive. Its functions are advisory (general advice to farmers), administrative (administration of farm improvement and other schemes), and educational (education of farmers in agricultural matters). All these functions have, to a greater or lesser extent, some bearing on pollution control. Agricultural inspectors, generally appointed by the local appointment board (a centralised body), are responsible for the administration of the various schemes of their committees.

The committees work very closely with the Department of Agriculture and Fisheries, which provides roughly fifty per cent of their finance. The Agricultural Inspectorate, which is part of the Department, supervises their work and provides them with advice and up-to-date information on matters pertaining to agriculture. The annual estimates are sent to the department for sanction as are the minutes of the monthly meetings held by each committee.

## 1.4.6 Boards of fishery conservators

The Fisheries (Consolidation) Acts, 1959 to 1964, provided that certain functions be exercised by boards of conservators, in seventeen fishery districts specified in the Act.[88] Boards consist of members elected quinquennially, *ex officio* members and holders of certain licences.[89] The boards' functions include protecting fish life generally and for this purpose they have certain powers to prosecute for pollution.[90]

Boards are responsible to the Minister for Agriculture and Fisheries[91] who has, *inter alia,* power to dissolve a board by order if satisfied that the duties of such board are not being duly and effectually discharged.[92] Annual accounts are audited by an auditor appointed by the Minister for Local Government.[93]

## 1.4.7 Harbour authorities

Harbour authorities were established by the Harbours Act, 1946, 1947, and are responsible for the administration of the harbours legislation in the commercial harbours scheduled in the 1946 Act. Each harbour authority is a body corporate. Membership consists of representatives of local authorities, users of the harbour, and labour interests, some members being nominated by the Minister for Transport and Power. Their functions in relation to pollution control are exercisable under the Petroleum Acts, 1871-1881, the

Harbours Act 1946, 1947, and the Oil Pollution of the Sea Acts 1956, 1965.

Harbour authorities work under the general supervision of the Minister for Transport and Power, who exercises controlling powers over them somewhat similar to, though less extensive than, the powers exercised over local authorities by the Minister for Local Government.

## 1.5 INDEPENDENT ADVISORY BODIES

Independent advisory bodies with pollution control functions are few, and none is specifically or exclusively charged with controlling pollution. Those which are particularly relevant, however, are

1. An Foras Forbartha
2. The Institute for Industrial Research and Standards
3. The Industrial Development Authority
4. An Foras Taluntais
5. The Inland Fisheries Trust Incorporated

### 1.5.1 An Foras Forbartha

An Foras Forbartha (The National Institute for Physical Planning and Research) is a state-sponsored body, established in 1964 by the Minister for Local Government to provide advice and undertake research in physical planning and development, in building and construction and in roads and water resources. It is responsible to, and works very closely with, the Minister for Local Government, and it derives its income mainly from state grants and charges for services. It is organised in four divisions, each concerned with research, training and advisory work covering the four

**1.5.1 continued** main aspects of the physical environment. The planning division deals with urban and regional planning and amenity and nature conservation, and publishes information and advice on these matters, which is available to planning authorities. The roads division deals with road traffic, road construction and road safety. The construction division deals with productivity, design and efficiency in the construction industry. The water resources division was established to 'collect process and evaluate information on water resources in the state, both as to distribution and availability of resources and their quality, with particular emphasis on pollution aspects. Pursuant to its functions in relation to water quality, this division, in cooperation with other bodies concerned, processes and collates hydrometric data on all river catchments; it establishes gaps in data and arranges with relevant bodies for further surveys and analyses, and stores and publishes the relevant information.'[94] An Foras Forbartha is mainly concerned with applied research to solve key issues in public environmental policy and much of its work follows requests from the Minister for Local Government and local authorities in respect of environmental issues which require advice and research by experts. Over two-thirds of its current research projects were requested by government departments, local authorities or other public bodies.[95] The role of An Foras Forbartha is to be 'an environmental watchdog, free of commitment to any polluting interest and to recommend necessary action, including new legislation, to deal with pollution, either generally or in particular cases.'[96] It has, in fact, contributed greatly towards dealing with pollution in Ireland and to this end it has, *inter alia,* published a vast number of valuable and ostensibly objective reports on general and particular aspects of the environment.[97]

An Foras Forbartha, when requested, advises planning authorities and other bodies concerned with pollution

1.5.1 continued

control on the qualitative and quantiative standards which ought to be imposed for controlling pollution. These standards may become mandatory if incorporated in legally binding pollution controls. It is particularly expert in, and concerned with, water quality aspects of effluent disposal and with the location of industries. It has a close working relationship with the Department of Local Government, other government departments concerned with pollution control, local authorities, other bodies mentioned in this section and with Bord Fáilte.

## 1.5.2 The Institute of Industrial Research and Standards

The Institute of Industrial Research and Standards is a state-sponsored body, responsible to the Minister for Industry and Commerce. Notwithstanding that it has no statutory functions in the field of pollution control, a large proportion of its activities are concerned with pollution problems. It provides the following services:

(i) Field advisory service — by which technical services are provided to industries concerned with pollution;

(ii) Process development service — designing and developing pollution-control processes for industries;

(iii) Equipment development service;

(iv) Laboratory service — it has three laboratories dealing with air and water pollution;

(v) Standards advisory service — these services were developed primarily for the Industrial Development Authority, but are available on request to all bodies dealing with pollution.

The IIRS is particularly concerned with problems relating to pollution by industrial effluent and its role

**1.5.2** **continued** embraces the carrying out of 'research, advisory and testing work for industrial organisations or for others when requested on methods of treatment of industrial effluents and the siting of factories in relation to industrial effluent, and to advise, as required, the Industrial Development Authority in this matter with regard to the siting of new factories, and to carry out research and development with regard to the utilisation of industrial wastes.'[98] The IIRS has published standards for effluent treatment which are widely used by pollution control authorities.[99] It numbers among its clients a large number of Irish and international companies, government departments, state-sponsored bodies, local authorities and others.[100] It has a particularly close relationship with the IDA — in the two years to 31 July 1973, over two hundred and fifty grant applications to the IDA were examined and assessed by the IIRS from the standpoint of water pollution control. In each case, recommendations concerning treatment standards and methods were made and followed through with the successful applicants. The IIRS and the IDA have an agreement by virtue of which the IIRS:

1. assesses the nature, quality and quantity of all effluent waters from proposed new industry or from proposed expansion of existing industry,

2. collects from the various knowledgeable bodies all relevant information on receiving water or local sewage systems, and, in the light of this, recommends treatment standards for the effluent according to the guidelines set out in S.R.5: 1970,

3. performs a technical audit on the ability of the proposed treatment plant to bring effluent quality to the recommended standard,

4. carries out performance tests where necessary to ensure compliance with standards imposed, and

**1.5.2 continued** 5. liaises with local authorities, industrialists, architects and engineers to satisfy, as far as possible, the requirements of these groups.

The IIRS also advises local authorities on suitable standards in connection with particular planning applications. The implementation of this advice varies from authority to authority. It has been consulted by prosecution authorities when prosecuting for pollution and when determining the acceptability or otherwise of the 'best practicable means' as a defence.

The standards set by the IIRS are not in themselves mandatory, but they may become so if adopted as conditions for licences, financial aid or for the grant of a planning permission.

The role of the IIRS in the field of pollution control is, however, suspect. It derives a substantial proportion of its income from private industry, [101] and there appears to be a potential, if not an actual, conflict of interests in this body which advises both controlling authorities and those who are likely to cause pollution.

## 1.5.3 Industrial Development Authority

The Industrial Development Authority was founded in 1949 as an agency of the Department of Industry and Commerce and became an autonomous state-sponsored body in 1970. It is responsible to the Minister for Industry and Commerce. The IDA is the main national body charged with promoting industrial development in Ireland. In this role it attracts, informs, assists and liaises with the industries it encourages. In order to attract and develop industry, the IDA has wide statutory powers to give financial and other assistance to qualifying industries. The amount of these grants may vary considerably with the geographical location of the grant-seeking industry —

**1.5.3** **continued** industries in areas designated for industrial expansion being considered more worthy of encouragement.[102] Grants may be made to qualifying industries on such terms and conditions as the IDA shall consider proper towards e.g.

(i) the cost of fixed assets,[103]

(ii) the cost of leasing fixed assets,[104]

(iii) towards the reduction of interest payable on loans raised to provide fixed assets,[105]

(iv) towards the cost of purchasing and leasing fixed assets required for the re-equipment, modernisation, improvement or expansion of existing industry,[106]

(v) for the training of persons in the processes of an industrial undertaking,[107]

(vi) towards the cost of research and development which the IDA is empowered to encourage,[108]

(vii) to secure the reduction of factory rents.[109]

The IDA may also guarantee the repayment of loans,[110] lease land at a concessionary rent,[111] and construct and administer industrial estates and advance factories (other than at Shannon and in the mid-west region).[112]

The IDA has not always been aware of its responsibility for the environment in fulfilling its role, but the contents of a recent policy statement, in which it was said that 'it is the policy of the IDA to discourage industries which are likely to cause pollution of the environment by effluent discharges and/or air emissions, and to require that all new industries and expanding industries comply with the conditions laid down by the controlling authorities,'[113] are

**1.5.3 continued** encouraging. Because of its ability to give aid to industry, especially substantial financial aid — new industries generally seek the assistance of the IDA. The IDA has an established procedure for dealing with such industries, viz. as soon as a serious level of negotiation has been reached between the IDA and the new industry, the latter is required to provide full details of any industrial effluent discharges or air emission from its proposed plants. These details are sent to the IIRS and to An Foras Forbartha and are examined by one or both of these bodies which return recommendations to the IDA setting-out measures necessary to keep any pollution at an acceptable level. These recommendations are forwarded by the IDA to relevant control authorities in order to assist them in framing suitable conditions in planning permissions and licences and in an attempt to maintain some consistency between the requirements of different control authorities. This practice, though it may be necessary and convenient, could possibly be a dangerous one when one considers the inexpertise of most planning authorities. Financial aid is given to the prospective industries subject to compliance with conditions attaching to planning permissions and the obtaining of any necessary licences. Non-compliance with a term or condition attached to a grant or other payment results in the amount of such grant or payment becoming repayable to the authority and, in default of being repaid, it is recoverable as a simple contract debt.[114] The grant or withholding of IDA aid to industry could be an effective weapon in persuading an industry to solve its pollution problems.

The IDA itself has no laboratory facilities but it uses the services of other bodies, notably the IIRS and An Foras Forbartha.

The power of the IDA to control industrial pollution is considerable. In the financial year 1973-1974 £23.1 million was pledged to domestic industry of which £12.6 million

**1.5.3 continued** was granted to modernise, re-equip and expand existing industries. Nevertheless, the IDA record for using its powers to control pollution may validly be criticised. Some of the industries which received substantial aid from this body are major contributors to the pollution problem in Ireland — a fact which encourages a certain amount of scepticism with regard to pollution control standards set by the IDA. In spite of its recently declared policy not to encourage industries likely to cause pollution, a number of environmentally undesirable industries propose to establish themselves in the country, especially in the south-west. It may be difficult for the IDA to impose stringent pollution-control standards because its main function is to establish industry in the country and it is on its success in this field, rather than in the field of environmental control, that it is judged.

## 1.5.4 An Foras Taluntais

An Foras Taluntais (The Agricultural Institute) was established under the Agriculture (An Foras Taluntais) Act, 1958. Its functions include reviewing, facilitating, encouraging, assisting, coordinating, promoting and undertaking agricultural research in Ireland. It is responsible to the Minister for Agriculture and Fisheries, to whom it submits its annual report. It has no specific statutory function in the field of pollution control, but its work contributes to the control of pollution, especially agricultural pollution. It is specifically obliged to disseminate the results of agricultural research, particularly to those engaged in agricultural advisory work, and it has a statutory obligation to advise the Minister for Agriculture and Fisheries on any matter relating to agricultural research or agricultural science on which advice is requested by him. Special sections in the Institute are concerned with pollution, viz. the farm structures and environment department, the soil fertility and chemistry department, the pesticide residue and analytical services unit, soil physics,

**1.5.4 continued** field investigations, soil biology and animal husbandry departments, the dairy microbiology department and the analytical services laboratory. The Institute has given high priority to pollution matter and has, *inter alia,* reported on pollution in Irish lakes; began a major study of animal manure disposal; surveyed fertiliser use practice; carried out experiments to measure surface run-off and infiltration through the soil of effluent from silage bins, monitored pesticide residues in soil, crops, foodstuffs and animal tissue, and carried out testing of chemical formulations. Qualitative and quantitative standards recommended by the Institute, while not legally binding, may be incorporated in legislative and other measures designed to control pollution. The role of the Institute extends to the evaluation of the effect, under Irish conditions, of various standard levels recommended by international bodies or used in other countries, and reporting thereon.

The Institute has a number of liaison committees which maintain links with the Department of Agriculture and Fisheries and other state organisations, notably An Foras Forbartha, the IDA, the IIRS, and Bord na Mona. Information is made available on request to sanitary, planning and other authorities, and the Institute has been involved with a number of local authorities in relation to planning applications for large industrial units where it was thought that pollution from the proposed industry could have an unfavourable effect on agriculture in the vicinity. The staff of the Institute are often involved in *ad hoc* working groups and are consulted by various government departments when information on specific subjects is required.

The Agricultural Institute is comparatively well equipped, well financed and well staffed. It is not subject to the same conflicting interests as is, e.g. the IDA, and only 2.7% of its finance comes from private industry. Its role as a pollution control authority should not be under-estimated because controls over pollution by, and of, agriculture are largely extra-legal.

## 1.5.6 The Inland Fisheries Trust Incorporated

The Inland Fisheries Trust Incorporated is a registered company, set up by the government in 1951, originally for the development of brown trout fishing, but from 1957 its articles were amended to provide for the development and promotion of all forms of angling as a national amenity and tourist attraction. The Trust is governed by a council of seven members, four of whom, including the chairman, are nominated by the Minister, and the remaining members are elected every three years by members of the Trust. The Trust itself owns substantial waters, and can, in certain circumstances, develop other waters. It has a fulltime staff of one hundred and fifty people, including pollution-prevention officers. It works in close cooperation with the Fisheries Division of the Department of Agriculture, with Boards of Conservators and with Bord Failte. Its income is provided by the state and by membership fees. Its functions in relation to pollution control arise by virtue of its ownership of fishery waters.

## 1.6.1 Special interest groups

There are an enormous number of organisations and bodies concerned with preventing pollution in Ireland. Some of these, though not necessarily the most important or the most effective, are treated in this section.

### (i) An Taisce

An Taisce—the National Trust for Ireland—is a voluntary non-profitmaking company, financed by private donations and subscriptions from members. It is concerned with the whole environment of the country and its principal aim is to conserve and develop the nation's physical heritage. It is a 'prescribed authority' for the purposes of S 21(1)(a) of

**1.6.1 continued** the Planning Act,[115] and planning authorities are obliged to send it copies of proposed development plans, or of proposed variations of such plans, and to take its objections and representation into consideration before making conservation orders. It has no other right under pollution control legislation, but it is vigilant and active in availing of any rights available to the individual.[116] It is perhaps the most active, and certainly the most vociferous, body concerned with pollution in the country.

**(ii) Bord Failte Eireann**

Bord Failte is a corporate body established under statute and financed by annual government grant. It is concerned with all aspects of tourist development, including the development of amenities for which it maintains a special fund. It is a prescribed authority for the purposes of S 21(1)(a) of the Planning Act and is consulted by planning authorities when making conservation orders. It has been said that Bord Failte 'appears to be the only body with sufficient financial and manpower resources to be able to meet satisfactorily the obligations placed upon it by the 1963 Planning and Development Act.'[117] The board has expressed itself as being 'extremely concerned with the increasing pollution of our waters' and has called for remedial action.[118] It has organised 'Tidy Town Competitions' which result to some extent in the alleviation of visible pollution, and it commissioned a study on the national coastline which recommended areas for conservation and indicated areas suitable for development.[119] The board is naturally extremely concerned with the ill-effects of pollution on the tourist trade.

(iii) **The Electricity Supply Board**

The Electricity Supply Board is a statutory corporation which has a virtual monopoly for supplying electricity in Ireland. It has eleven generating stations operating on peat, eight on oil and coal, and nine hydro stations. It is exempted, for some purposes, from the provisions of the

**1.6.1 continued** Planning Act, 1963,[120] but has voluntarily accepted the provisions of the Act in relation to transmission line routing for all lines over 38 KV in all areas, and for lines of lower voltage in areas of high amenity value. It owns extensive fisheries in its own right and has powers of control over these. In addition, the Electricity Supply (Amendment) Act, 1945, S 42, gives the ESB a measure of control overall rivers and streams serving electricity generating stations. It prohibits any person, without the written permission of the board, from discharging or allowing into a river (or into any tributary of any river, or any watercourse connected with any river) which is to be used by the board in connection with the generation of electricity, any chemical or other substance which might injure any part of the generating station or any works subsidiary to it or connected with it. Contravention of the provision is punishable by a £50 fine plus £20 for every day on which the offence is continued. Prosecutions have been taken in a number of cases. The ESB spent £161,000 in the financial year 1972-1973 in the preservation and development of amenity resources under its control and on fisheries.[121] The board employs an administrative officer in its head office and two pollution investigation officers in the field. Its fisheries protection staff are instructed to report on cases of pollution which come to their notice in their daily rounds. It has its own central laboratory for investigating, *inter alia,* pollution, together with facilities for BOD, pH and ammonia analyses. When necessary, it avails of the expertise of the public analyst and other bodies, e.g. An Foras Forbartha, the IIRS and An Foras Taluntais. It has regular consultations with local authorities in their various capacities and is represented on various planning committees, regional development organisations, on many boards of conservators and on the General Council of the Board of Conservators. It operates a close liaison with fisheries interests especially with the Inland Fisheries Trust and the Department of Agriculture and Fisheries. It monitors air pollution in a number of locations, and has its

**1.6.1 continued** own internal standards for emissions. It has called for stricter laws to control pollution,[122] and, on the whole, it would appear that the board is aware of its responsibilities with regard to the environment.

(iv) **The Keep Ireland Beautiful Movement**
The Keep Ireland Beautiful Movement was founded by a group of business leaders and industrialists in 1972 under the auspices of the Confederation of Irish Industry. Its overall view is 'to develop among business, industry and the general public, an awareness of the need to preserve and protect the natural beauty of Ireland from the growing menace of litter and dumping.' It seeks to attain its objectives by reporting on the deterioration of the environment, by informing the public of the need to be watchful, by liaison with interested organisations, by encouraging waste recycling, and by providing the necessary services to attain these objectives, and in particular, youth education programmes. It is, as yet, difficult to evaluate the achievements of this movement.

(v) **Others**
Other associations concerned with preventing pollution include the Inland Waterways Association of Ireland, the Maritime Institute, local field clubs, residents' associations and local development committees.

## 1.6.2 Individuals

The remedies available to an individual adversely affected by pollution vary and overlap between those offered by the Constitution,[222] by the courts at common law and in equity, and those offered by statute, by subordinate legislation and by EEC law. These paragraphs propose to deal with the more general individual rights—specific rights of one type or another will be treated at later stages of this report. The common law remedies available to an individual are founded principally on actions in tort, i.e. trespass, nuisance, negligence and under the rule in *Rylands* v. *Fletcher*. Rights and remedies are also conferred on the individual by statute.

## 1.6.2.a The Constitution

See 1.2.1.

## 1.6.2.b Trespass

Pollution may, amongst other things, constitute a trespass. A trespass may be committed by placing or projecting any material object on land or causing any physical or noxious substance [124] to cross the boundary of the plaintiff's land or even simply to come into physical contact with the land, though there may be no crossing of the boundary. [125] The injury must be direct in order to be actionable in trespass. The remedies available to the successful plaintiff are damages and/or injunction.

## 1.6.2.c Nuisance

**Public nuisance**

Nuisance may be either public or private, or both. A public nuisance is a crime, but it is also a tort for which an individual may, in the right circumstances, be able to recover damages or obtain an injunction. Public nuisance must affect a number of people in a given area, and each of them must be so affected. In order to be actionable by an individual, the public nuisance must be such that individual suffered damage over and above that suffered by the public at large.

**Private nuisance**

Private nuisance usually consists of an 'act of wrongfully causing or allowing the escape of deleterious things into another person's land — for example, water, smoke, smelly fumes, gas, noise, heat, vibrations, disease-germs, animals

**1.6.2.c continued** and vegetation.'[126] The escapes must generally cause damage to land or things thereon or to the occupier of land. The escape must, as a rule, have occurred over a substantial length of time. Generally, nuisance is occasioned by the 'unreasonable and unnecessary' use of property by a defendant in such a way as to inconvenience his neighbour, and the court will consider, *inter alia,* the type and gravity of the harm caused by the escape and the area in which the alleged nuisance occurred. The action is maintainable at the suit of an individual who is either an occupier of land (usually the owner or lawful tenant, but not a licensee), or a person who has suffered permanent damage to his reversionary interest. An exceptionally sensitive plaintiff cannot expect a remedy for a nuisance which would not affect others less sensitive.[127] The requirements which must be met for a successful action based on private nuisance as applicable to cases of environmental pollution result in certain deficiencies in the protection afforded to the individual. To some extent, these deficiencies have been remedied by the legislature.[128] The remedies available to a successful plaintiff in an action for nuisance are damages and/or an injunction. In addition, a person affected by a nuisance may, in certain circumstances, take steps to abate the nuisance.

**Statutory nuisance**

Public health legislation provides that certain activities shall constitute statutory nuisances. Most of the statutory nuisances which are relevant to pollution control are contained in the Public Health (Ireland) Act, 1878. Section 109 of the Act provides that an individual aggrieved by a nuisance under that Act, or any two householders of a sanitary district may inform a sanitary authority of existence of the nuisance, must take action, direct or indirect, to remove it.[129] An individual may also complain to the courts of a nuisance under the said Act.[130] The above provisions are in addition to, and do not affect, any other remedies for pollution which may be available to the individual.[131]

### 1.6.2.d Negligence

Negligence consists of a breach of a legal duty of care owned by one person to another as a result of which injury is caused to the other. It is, *prima facie,* a remedy for an individual who has suffered damage from pollution of any type.

### 1.6.2.e *Rylands* v. *Fletcher*

The rule in *Rylands* v. *Fletcher* imposes absolute or strict liability upon the occupier of land who brings and keeps upon it anything likely to do damage if it escapes. Such an occupier is bound to prevent its escape and is strictly liable for all the direct consequences of its escape. This remedy is, or can be, available in respect of a wide variety of pollutants which escape, but is not available for an escape resulting from a 'natural' use of land. The action, unlike an action in nuisance, may be brought by anyone who has suffered damage.

### 1.6.2.f The Planning Act 1963

The individual has certain rights under the Local Government (Planning and Development) Act 1963, and regulations made thereunder by virtue of which he can endeavour to prevent or minimise the pollution. *Any* individual has the right to inspect a copy of a draft development plan, and to object to, or make representations with respect to, that draft and, if he is a ratepayer, he may include in his objection a request to be afforded an opportunity to state his case before a person or persons appointed by the planning authority, which objection and opportunity shall, respectively, be considered and afforded.[132] *Any* person may furnish written objections to the Minister for Local Government within twenty-one days of receipt by the said minister of an application by a planning authority for his consent to allow them to grant a

**1.6.2.f continued** permission for development which would contravene materially any development plan or any special amenity order relating to their area, and the Minister is bound to consider such objections before he grants his consent.[133] *Any* person may appeal to the Minister against a decision of a planning authority to permit a development or the retention of a structure provided he does so within the prescribed period, i.e. one month from the day of receipt of the decision where the appellant is the applicant, and twenty-one days from the date of the decision in any other case.[134] In addition, any person may appeal to the Minister in relation to the granting, refusing, withdrawing or continuing of a licence under S 89 of the Act or to the conditions specified by the planning authority for such licence.[135] It has been held that there is nothing in the Act to suggest that an applicant for permission must have an estate or other interest in land,[136] and the same probably holds true for any person objecting to the granting of a permission. It is obviously unfair, however, that the time within which appeals must be made is shorter for an appellant who is not the applicant than it is for the applicant himself, who is, of course, in full possession of all the facts and who has ample time to prepare arguments where he anticipates an appeal. Regulations made under the Act[137] provide for the procedure in relation to appeals and references to the Minister. These regulations provide, *inter alia,* that any party to an appeal may request, and the Minister may direct, an oral hearing of the issue. The right to request an oral hearing has frequently been used by An Taisce and other bodies seeking to control pollution. The hearing is conducted by an inspector appointed by the Minister. The function of the inspector is that of a 'recorder for the Minster' and it has been held that 'he has no advisory function, nor has he any function to arrive at a preliminary judgment... or to arrive at any conclusion'.[138] The regulations empower the inspector, if he thinks fit, to hear any person not party to the appeal — hence another opportunity of airing their views for those interested in

**1.6.2.f continued** preventing pollution. It has been held that the inspector has an implied right to consider matter not put before the planning authority.[139] The inspector reports on the hearing to the Minister. The Act, S 26(5)(b), provides that the Minister, when considering an appeal, 'shall determine the application as if it had been made to him in the first instance, and his decision shall operate to annul the decision of the planning authority as and from the time when it was given.' This sub-section has been judicially interpreted to mean that 'the Minister may impose conditions not imposed by the planning authority, touching perhaps on matters not even dealt with by the planning authority' and that 'S 26(5)(b) does not confine the Minister to considering the identical application that was before the planning authority provided... that any alteration in that application made between the decision of the planning authority and the hearing of the appeal is known to the objectors.'[140] In the Killiney and Ballybrack case (an application to the High Court brought by a residents' association, motivated by a desire to prevent pollution) it was held that 'in reaching any decision with regard to the question of sewage or pollution as part of a decision to grant or refuse planning permission for this development, the Minister would be concerned not only with the existing situation, with regard to sewage facilities, but with the probable position for future development or expansion of any sewage facilities and would be concerned primarily to decide whether, if and when the development became occupied, sewage facilities sufficient to dispose of the sewage from it would be available, and pollution arising from it would thus be avoided.' The Minister would, therefore, seem to have wide powers to consider the potentiality of pollution when making decisions on appeal. The aforementioned individual rights are valuable weapons for drawing attention to, and preventing, pollution, and have been so used.

## 1.7.1 National policies on pollution control

Until recently, Irish Governments have been relatively unconcerned about pollution control as witness the paucity and antiquity of legislative controls of pollution. To some extent, this attitude was justified as industrial and agricultural expansion, the result of which now causes severe pollution problems, are recent phenomena in Ireland. Where pollution could be foreseen in the past, the tendency was, and to some extent still is, to give priority to the necessity to cure a chronic and persistent unemployment problem even when doing so would result in polluting the environment.

In so far as it can be ascertained, national policy is to treat pollution control as a local or perhaps a regional matter to be dealt with on a piecemeal basis — hence, the major pollution control authorities are local authorities. Responsibility for pollution control lies primarily with local politicians, with some but by no means any extensive control lying with central government. There is no evidence of an inclination to depart from this policy, and repeated calls to establish a government department with sole and exclusive responsibility for the environment have been ignored.

Existing legislation designed to prevent or minimise pollution is now generally recognised as being inadequate, yet the customary practice of copying British legislation in most matters has not been followed in respect of pollution control. The Local Government (Planning and Development) Bill, 1973,[141] is hardly evidence of a strong determination to strengthen the anaemic control powers of planning authorities. A new Bill, now at the drafting stage, may provide much needed reforms.

## 1.7.2 Types of control used

The main types of pollution controls used are:

1. Refusal to grant planning permissions or approvals, or the attachment of conditions to same;

2. Individual rights to intefere in the planning process;

3. Land-zoning contained in development plans drawn up by planning authorities;

4. Operating conditions in licences, leases and permissions;

5. Withdrawal or refusal of grant-aids unless pollution control measures are taken or complied with;

6. Approval of products before they are marketed or recommended for use;

7. Ministerial circulars to pollution control authorities;

8. Common law actions;

9. Public opinion.

## 1.7.3 Other policies

There has been much concern at central and local levels with identifying and quantifying pollution. Committees set up by the Government have reported on water[142] and oil[143] pollution and the implementation of recommended control is being considered. An Foras Forbartha, at the request of the Minister for Local Government, have carried out a national survey of Irish rivers.[144] There have been many investigations of pollution on a less extensive scale.

**1.7.3 continued** Three state-sponsored bodies have been established to carry out research and provide services in, *inter alia,* the field of pollution control, viz. An Foras Forbartha, the Institute of Industrial Research and Standards, and An Foras Taluntais.[145]

Ireland is participating in the European Project on Pollution and has signed an agreement relating to the project dealing with the analysis of micropollutants in water as well as indicating its willingness to participate in four other major projects.[146]

The position with regard to responsibility for the cost of clearing up pollution is being reviewed. Present legislation merely provides set penalties for pollution, and these bear no relation to the cost of taking remedial action after pollution has been caused.

1. Professor B. Chubb, *A Sourcebook of Irish Government,* pp. 156-157.
2. See also 1.4.1.
3. Para. 2.2.1. Local Government Re-organisation Prl. 1572 (1971).
4. Para 3.3.5. Report of the Committee on Public Services Organisation. Prl. 792.
5. See also 1.4.1.
6. Sean F. Lemass (former Taoiseach), *Administration* Vol. 6.
7. Para. 5.2.6. E.P.S.O.R.G. supra.
8. Usually public utilities, bodies with strong social leanings and those with promotional and/or research functions.
9. The more commercial types.
10. Sean F. Lemass *Administration* 'vol. 6.
11. See Dail Debates, Vol. 182, cols 999-1003.
12. *Ryan* v. *A.G.* (1965) I.R. at p.344 ( S.C.).
13. Id., p.345.
14. Public Health (Ireland) Act 1878 S 107. This section has been repealed and replaced in so far as it relates to offices and factories by the factories Act 1955 and the Office Premises Act 1958 respectively.
15. Local Government (Sanitary Services) Act 1948 S 18.
16. Id., S 32.
17. Local Government (Sanitary Services) Act 1962 S 10.
18. See *Street on Local Government* p.146.
19. *Banbury Urban Sanitary Authority* v *Page* 8 Q.B.D. 97.

20. See Report on Water Pollution 1972. Prl2939. The committee has not yet reported on air pollution.
21. Local Government Re-organisation Prl. 1572.
22. Except the county of Tipperary which has two county councils.
23. See Appendix I.
24. Local Government Re-organisation Prl. 1572.
25. Id., para 1.3.2.
26. See 1-4-5.
27. Local Government Re-organisation para 2.2.1.
28. See Central—Local Relationships in Ireland, *Administration* Vol. 3 pp.110-111.
29. S 59 Local Government Act, 1955.
30. Local Government Services (Corporate Bodies) Act 1971.
31. See *Irish Times* 23 October 1973.
32. See Report on Water Pollution Prl. 2939, para 7.15.1.
33. See Local Government Re-organisation. Prl. 1572 (1971) and Strengthening the Local Government Service Prl. 2252 (19717.
34. Local Government (Planning and Development) Act 1963 S 19 and Third Schedule.
35. Id., S 19 and L.G. ( & D) Act 1963 (Copies of Development Plans) Regulations, 1967 S.I. No. 154/1967.
36. Id., S 22.
37. Id., S 23.
38. Id., S 22.
39. Id., S 24 and L.G.(P&D) Act, 1963 (Appointed Day) Order 1964. The appointed day was 1 October 1964.
40. Id., S 3.
41. Id., S 4, and L.G.(P&D) Act, 1963 (Exempted Development) Regulations SI No. 176 of 1967 and SI No. 260 of 1968.
42. Id., SS 10/25, and L:G. (P&D) Act 1963, (Permission) Regulations S.I No. 221 of 1964.
43. L.G. (P&D) Act, 1963 (Permission) Regulations 1964 Reg. 9.
44. L.G. (P&D) Act, 1963, SS 36, 27, 28.
45. Id., S 26(i) and S 27(i).
46. Id., S 89 and L.G.(P&D) Act, 1963 (Licensing) Regulations S.I No. 76 of 1965, S.I No. 72 of 1968, S.I No. 210 of 1968.
47. Id., SS 86, 87, 88.
48. Id., S 28 (5).
49. Id., S 24.
50. Id., SS 31(8), 34, 35(7) and (8), 37(7), 45(8), 46, 52(4), 53(2), 49(2), 51(4), 86(8), 89(7).
51. Id., S 80.
52. Id., SS 30, 39.
53. Id., SS 30, 39.
53. Id., SS 31, 32, 35.
54. Id., S 31.
55. Id., S 32.
56. Id., S 35.
57. Id., S 36.
58. Id., SS 36, 29.
59. Id., S 37

60. Id., S 2.
61. Id., S 81.
62. Id., S 44
63. Id., SS 42, 43.
64. Id., S 45
65. Id., S 46.
66. Id., S 50.
67. Id., S 54.
68. Id., S 77.
69. Id., see especially S 16, S 23(2), and S 42(3).
70. See 1-5.
71. L.G. (P&D) Act, 1963, S 14.
72. Id., S 21 and S 46(7).
73. A.C.R.A. Sub-Committee Report on Planning Laws 1972.
74. S 4 L.G. (P&D) Act, 1963, SI 176 of 1967 and SI 260 of 1968.
75. L.G., (P&D) Act, 1963, S 92, Housing Act, 1969, S 9.
76. L.G, (P&D) Act, 1963, SS 26(4), 27(3); Housing Act 1969, S 4(5).
77. L.G. (P&D) Act, 1963, P vi and L.G. (P&D) Act, 1963, (Compensation) Regs. SI No. 217 of 1964.
78. Id., S 29
79 Appendix 6.
80. See, e.g. The Lower Shannon Pollution Investigation, *Water Resources in the North East* and *Water Resources Survey in County Donegal* Commissioned by the Mid-Western, North-Eastern and Northern R.D.O.'s respectively.
81. See Public Health Act, 1878, S 219, 1-2-3, and 3-1.
82. Report on Water Pollution, Prl. 2939, para. 5-4-2.
83. See 3-1.
84. See 2-1.
85. Report on Water Pollution, para. 10-2-1.
86. Dail Reports, Vol. 261, Col. 1853 (20 June 1972).
87. For a criticism of this system see Report of the Public Services Organisation Review Group, Prl. 792, para. 26-2-11 and 26-2-13.
8o8. Second Schedule Fisheries (Consolidation) Act, 1959.
89. Fisheries (Consolidation) Act, 1959, S 21.
90. Id., SS 164, 171, 172 (as amended by S 18 Fisheries (Amendment) Act, 1963).
91. Fisheries (Transfer of Departmental Administration and Ministerial Functions) Order, SI, No. 83 of 1965.
92. Fisheries (Consolidation) Act 1959, S 51. See e.g. SI No. 281/1973 dissolving the Cork Board.
93. Id., S 48.
94. Report on Water Pollution Prl. 2939, para. 7-11-1.
95. Forbairt, 1973.
96. Memorandum of Understanding between An Foras Forbartha and other bodies concerned with pollution.
97. See, e.g. *The Importance of Conservation in County Kerry* (1968), *Brittas Bay_A Case Study in Conservation* (1970), *Conservation_The Human Environment* (1971), *National Survey of Irish Rivers* (1972), *National Coastline Study*(1972).
98. Memorandum of Understanding between An Foras Forbartha, the IIRS and others.

99. S.R.5: 1970.
100. In 1971-'72, forty per cent of its clients were non-agricultural-based industries, thirty per cent agricutural-based industries, and the remainder consisted of semi-state bodies, local authorities and others.
101. In 1971-'72 seventy per cent of its clients were actual or potential polluters. See also Annual Report 1972-'73, pp.14, 40.
102. See e.g. Industrial Development Act, 1969, SS 33(2), 35(2), 38(2), 46(2).
103. Id., SS 33, 34.
104. Id., S 35.
105. Id., S 36
106. Id., S 38.
107. Id., S 39.
108. Id., S 40
104. Id., S 46.
110. Id., S 37.
111. Id., S 47.
112 Id., S 15.
113 IDA Policy Statement.
114. Industrial Development Act, 2969, S 45.
115. L.G. (P&D) Act, 1963, (Miscellaneous) Regulations, 1964, SI, No. 219/1964.
116. See 1-6-2.
117 *The Protection of the National Heritage,* 1969, p.24.
118 Annual Report 1973, p.19.
119 National Coastline Study, 1972.
120. S 4(1)(f) and L.G.(P&D) Act 1963 (Exempted Development) Regs., 1967, SI No. 176 of 2967, Classes 20 and 21, and L.G. (P&D) Act, 1963 (Exempted Development) '6amendment) Regs. SI No. 260 of 1968, Art. 2.
121 Annual Report 1973, p.15.
122 ESB Fisheries Report, 1973.
123. See 1-2-1.
124. *McDonald* v. *Associated Fuels* [1954] 3 D.L.R. 775 (blowing carbon monoxide into a house) and *Callan* v. *McAvinue* H.C. 11 May 1973 (unreported).
125. *Salmond on the Law of Torts,* 16th Ed., pp.41-42.
126. Id., at p.51.
127. *Mullin* v. *Hynes,* S.C., 13 November 1972.
128. See 1-2-3.
129. SS 110. 111.
130. S 121.
131. S 127.
132. Local Government (Planning and Development) Act, 1963, S 21.
133. Id., S 26(3).
134. Id., SS 26(5), 27(4).
135. Id., S 89(3).
136. *Frascati Estates Ltd.* v. *Walker,* H.C., 3 December 1973.
137. L.G. (P&D) Act, 1963 (Appeals and References) Regulations, 1964, SI No. 216 of 1964.
138. *Killiney and Ballybrack Development Association Ltd.* v. *Minister for Local Government and Templefin Estates Ltd.,* H.C., 1 March 1974.
139. Id.

140 Killiney and Ballbrack case *supra.*
141. See 1-4-2.
142. Report on Water Pollution, Prl. 2939.
143. Report on Oil Pollution, Prl. 1431.
144. *The National Survey of Irish Rivers_A Report on Water Quality,* 1972.
145. See 1-5-1, 1-5-2, 1-5-4.
146. Dail Reports, Vol. 263, col 975, (8 November 1972).

# 2 Air pollution

Air pollution in Ireland is not a serious problem outside the major cities and isolated industrial areas. Such pollution as exists is caused mainly by emissions from industry, from mechanically propelled vehicles and from domestic heating — the last mentioned being the major source of air pollution. It has been alleged that average levels for smoke and sulphur dioxide and the frequency of days of excess pollution in our two major cities are comparable to those of dense residential and industrial areas in the United Kingdom.[1] In spite of this, no legislation designed to control air pollution from domestic sources has been enacted, and there are no 'smokeless zones' in Ireland. The Control of Atmospheric Pollution Regulations 1970, which are the only legislative controls of air pollution enacted since the foundation of the state, are inadequate and anaemic and deal only with certain types of pollution from some industries and public places. The Alkali, etc. Works Regulation Act, 1906 is partly obsolete and has never been extended, as has its British counterpart, to cover new processes. The Department of Local Government records air pollution in the ambient air in forty-seven monitoring stations located in the major cities and towns. Many of these stations are of a most primitive type. Statistics on air pollution as a source of material damage do not exist. There are very few enactments which specifically empower control authorities to control air pollution. The legal basis for such control as is exercised springs largely from widely framed legislative clauses empowering control authorities to attach conditions as they think fit to permissions, licenses and leases. This is obviously an unsatisfactory method for controlling air pollution because in some cases the legal validity of conditions designed to control such pollution is questionable, and, in all cases, it would be far more effective if powers to control air pollution were specifically enumerated amongst conditions which might validly be attached to permissions, etc. Such specific enumeration would resolve ambiguities, would focus the attention of control authorities on the possibility of air pollution and would encourage applicants to provide satisfactory measures for preventing air pollution by their undertakings.

## 2.1.1 STATIONARY SOURCES

Controls over the siting of points of discharge of air pollutants are exercisable by planning authorities and the Minister of Local Government on appeal, through conditions attached to planning permissions and licences under the Planning Act;[2] by the Minister for Local Government who has wide powers to make regulations to control atmospheric pollution;[3] by the Minister for Industry and Commerce who is empowered, *inter alia,* to grant or refuse, at his discretion, licences to smelt ore and ore concentrates by way of trade or for the purposes of gain, and who may, if he decides to grant such licence, attach 'such conditions as he shall think proper' to the licence;[4] and to grant or refuse, at his discretion, a licence to manufacture cement,[5] and, if he decides to grant such licence, to attach to it certain specified conditions[6] which may result in the prevention or reduction of air pollution. The same Minister has wide legislative powers to control mineral workings and related activities. About sixty percent of all minerals are vested in the state. State minerals do not include those minerals in lease prior to 1940, but there are compulsory acquisition powers under the Minerals Development Act, 1940. Certain minerals are exempted from state control notably the bulk minerals — sand, gravel and limestone. The Minister may grant or refuse in respect of minerals prospecting licences,[7] licences to work state-acquired minerals,[8] subsidiary licences called mining facilities licences,[9] state mining leases,[10] mining facilities permits, mining facilities, (state land) permits,[11] state mining permissions,[12] unworked minerals licences,[13] and ancillary rights licences[14] broadly speaking, on such terms and conditions as he shall think proper and shall specify.[15] The Minister also has wide powers to control the operation and development of petroleum by virtue of the Petroleum

**2.1.1 continued** and Other Minerals Development Act, 1960. He may, in accordance with the provisions of that act, undertake to grant an exploration licence, a petroleum prospecting licence or both,[16] or a petroleum lease,[17] and may grant an exploration licence,[18] a petroleum prospecting licence,[19] a petroleum lease,[20] a working facilities permit, a working facilities (state land) permit,[21] and a reserved area licence,[22] broadly speaking, on such terms and conditions as he shall think proper and shall specify.[23] The Minister is also empowered to make regulations specifying, *inter alia,* the conditions to be complied with by an applicant for a state-acquired minerals licence or a mining facilities licence and specifying terms and conditions to be incorporated in such licences and obligations to be imposed by such licences on such licensees, and, in particular, requiring the licensee to have due regard to the amenities of the locality in which is situate the subject matter of such licence;[24] and for the siting and spacing of petroleum wells and such other measures in relation to petroleum operations as the Minister thinks necessary or expedient to provide for.[25] In practice, licences, etc., granted by the Minister, are granted subject to conditions designed to prevent pollution and generally specify that all operations be carried out in a proper and workmanlike manner in accordance with the methods and practices customarily used in good mining or oilfield practice and with the best practices of exploration, and that the grantee comply with any instructions from time to time given by the Minister relating to these matters.

The IDA may control the siting of points of discharge of air pollutants by conditions attached to grants of financial aid,[26] as may the Alkali Inspector when considering applications for registration or for the renewal of registrations of controlled works.[27]

**2.1.2** Controls over processes may be exercisable in the manner described by all the authorities mentioned in 2.1.1. The Local Government (Sanitary Services) Act, 1962, S 10,

**2.1.2**
**continued**

expressly empowers the Minister for Local Government to make regulations regulating *inter alia* 'the establishment and operation of

(i) trades,

(ii) chemical and other works, and

(iii) processes (including the disposal of waste),

which are potential sources of atmospheric pollution from smoke, dust, grit or noxious or offensive gases; licensing of persons engaged in specified works or processes, being works or processes discharging pollutants into the atmosphere, and prohibiting the engagement in such works or processes of persons other than licenced persons, and licensing of premises from which pollutants are discharged into the atmosphere, and prohibiting discharge of pollutants into the atmosphere from premises other than licenced premises.' Unfortunately, apart from making one set of regulations of limited effectiveness, the Minister has not used his powers under this section. Apart from the aforementioned, the principal legislation controlling processes causing atmospheric pollution is the Alkali, etc. Works Regulation Act, 1906. This Act requires that every alkali work,[28] scheduled work,[29] cement work or smelting work must be registered in accordance with the requirements of the Act before it can be carried on. Works controlled by the Act may be scheduled or non-scheduled. Control over the latter is somewhat less strict. It is a condition of first registration under the Act that the plant shall be equipped with the 'best practicable means' of preventing the discharge into the atmosphere of noxious and offensive gases and for rendering them harmless and inoffensive. 'Best practicable means' is interpreted as the provision, efficient maintenance, and proper use of appliances for preventing the escape of gases, smoke, grit and dust and for rendering them harmless and inoffensive

**2.1.2 continued** where discharged, and the proper supervision of operations causing such emissions. It takes into account the effect of such measures on the operation of the process and their cost since a balance has to be preserved between the cost and the extent of the harm or nuisance involved. The obligation to use the 'best practicable means' is a continuing one and may entail alterations in the plant and method as new techniques become available. Registered works are subject to inspection; the owner must provide facilities for inspectors to enter and inspect the premises and carry out tests, and he must disclose in confidence such details of the process as the inspector may require.[30] Apparatus for preventing pollution must be kept in good order and in continuous operation. In 1971, 1972, 1973 there were respectively twelve, eleven, eighteen scheduled works and two, two, three non-scheduled works registered under the Act. Some of the scheduled works involved more than one process, so that there were in all twenty-eight, twenty-five and twenty-five processes respectively controlled in the said years. Works producing sulphuric acid, chemical manure, nitric acid, sulphate of ammonia and tar were registered as scheduled works while all the non-scheduled works were cement works.[31] The Act has never been up-dated to include many modern processes and many of the processes specified in the Act are now obsolete. There is reason to believe that many processes which should be registered are not so registered. The Minister for Local Government has taken action, with no apparent success to remedy this situation.[32] Control over processes is exercisable by the attachment of conditions to certificates of registration, by threatening not to grant or renew a certificate, and by minimal penalties for pollution specified in the Act. Emissions from industrial processes not described herein may be subject to the control of local authorities under the nuisance provisions of Public Health Acts.[33]

**2.1.3** Requirements as to the pre-treatment of any discharge to air may be attached as conditions for the grant of permissions,

**2.1.3 continued** licences, leases and permits described in 2.1.1 and 2.1.2.

The Alkali, etc. Works Regulation Act, 1906, provides that the 'best practicable means' be used by the owner of every alkali work and scheduled work to prevent the escape and discharge into the atmosphere of noxious and offensive gases and for rendering such gases when discharged harmless and inoffensive. Failure to use such means is an offence punishable by fine.[34] The alkali inspector may inquire whether means can be adopted at a reasonable expense for preventing the discharge of noxious and offensive gas from the furnaces and chimneys of cement and smelting works, or for rendering such gas where discharged harmless or inoffensive.[35] Compliance with these provisions may necessitate pre-treatment of any discharge likely to cause atmospheric pollution.

**2.1.4** Limits on the composition concentration or quantity of discharge into the atmosphere may be imposed by the stated control authorites.[36] The Minister for Local Government has circularised planning authorities and asked them to consider attaching conditions to planning permissions in respect of emissions from fuel burning plants and emissions of grit and dust from chimneys.[37] The Control of Atmospheric Pollution Regulations, 1970, article 4,[38] makes it an offence for the occupier of any premises other than a private dwelling-house[39] to cause or permit to be emitted from such premises smoke, dust, grit, gas or fumes in such quantity or in such a manner as to be a nuisance to persons in any premises in the neighbourhood; but if the emission consists of dark smoke resulting solely from carrying out research relevant to the problem of air pollution no offence shall be committed provided that prior notice of the experiments causing the emission had been given to the sanitary authority of the district.[40] Where a person is charged with contravention of article 4, it is a defence to prove that the best practicable means have been taken to minimise the emission complained of. It is an

**2.1.4**
**continued**

offence under article 5 of the regulations for a person to cause or permit to be emitted from any public place smoke, dust, grit, gas or fumes in such quantity or in such a manner as to be a nuisance to persons in any premises in the neighbourhood and in this case it is not a defence to prove that the best practicable means were taken to minimise the emission. A public place means 'any street, road or other place to which the public have access with vehicles whether as of right or by permission and whether subject to or free of charge.'

The Alkali, etc. Works Regulation Act, 1906, prescribes maximum permissible limits for emissions from four processes only, viz. muratic acid gas in alkali works,[41] condensation of acid gases in sulphuric acid[42] and in muratic acid works[43] and acid gases in the air, smoke or gases discharged into the atmosphere from a chimney or other final outlet.[44] However, the Act further provides that no objection shall be taken by an inspector to any muratic acid gas from an alkali works in the air, smoke or gases discharged into the atmosphere by a chimney or other final outlet where the amount of such acid gas does not exceed the amount prescribed in respect of muratic acid gas evolved in alkali works,[45] nor to any muratic acid gas from scheduled works in the air, smoke or gases discharged into the atmosphere by a chimney or other outlet where the amount of such acid gas does not exceed the amount prescribed in respect of acid gases in sulphuric acid and muratic acid works.[46] Thus, apart from a few acid processes, specific values of acceptable emissions on ground level concentrations are not defined. Instead, the inspector uses 'presumptive limits' which, if not exceeded, will satisfy him that the best practicable means are being used. These limits are decided in the light of advancing technology and much attention is given to British practice in this field.

**2.1.5** Controls over the method of discharge may be exercised by the authorities mentioned in 2.1.1. Planning authorities do, in practice, specify conditions controlling the methods of discharge of air pollutants from industry. The Minister for Local Government drew their attention specifically to the necessity of controlling emissions from fuel-burning plants and emissions of grit and dust from chimneys and, for their guidance, referred them to, and furnished them with, memoranda on these topics published by the British Ministry of Housing and Local Government.[47] He also requested the authorities to furnish his department with conditions imposed by them in particular cases to control air pollution from fuel-burning plants.[48]

The Control of Atmospheric Pollution Regulations[49] set limits on the length of time during which dark or black smoke (that is, smoke as dark as, or darker than, shade 2 and shade 4, respectively on the Ringleman chart), may be emitted from premises other than private dwelling-houses. These limits may be shown in tabular form:

Article 3(1):

**Emissions from premises**

| *Emission* | *Maximum permissible time* |
|---|---|
| Dark smoke | 4 continuous minutes |
| Dark smoke | 8 minutes in the aggregate in any period of 8 consecutive hours |
| Black smoke | 2 minutes in the aggregate in any period of 30 minutes |

Article 3(2):

**Emissions from chimneys serving two or more furnaces**

Permitted period of dark smoke in any given eight-hour period:

**2.1.5** continued

| *No. of furnaces served by one chimney* | *Maximum permissible time* |
|---|---|
| 2 | 16 minutes |
| 3 | 22 minutes |
| 4 or more | 27 minutes |

Article 3(5):

**Defence to contravention of article 3(1)(b)**

If soot-blowing, permitted period of dark smoke in any given eight-hour period:

| *No. of furnaces served by one chimney* | *Maximum permissible time* |
|---|---|
| 1 | 12 minutes |
| 2 | 23 minutes |
| 3 | 32 minutes |
| 4 or more | 39 minutes |

A great number of defences are provided for those who exceed the prescribed time limits, viz. the defence that the best practicable means were used to avoid the contravention;[50] that the contravention was caused by the lighting up of the fire or furnace from being cold [51] that the contravention was solely due to some failure of a furnace or apparatus used in connection with a furnace; that the failure could not reasonably have been foreseen, or if foreseen, could not reasonably have been provided against; and, that the contravention could not reasonably have been prevented by action taken after the failure occurred.[52] Furthermore, when dark smoke is emitted from premises for a continuous period exceeding four minutes or for more than eight minutes in the aggregate in any eight-hour period, it is a defence to prove that the contravention was caused during the operation of soot-blowing,[53] but in the latter case, the emission must not exceed the time limits precribed by article 3(5) above. No offence is committed by

**2.1.5 continued** the emission of dark or black smoke from any premises resulting solely from the carrying out of research relevant to the problem of air pollution provided that prior notice of the experiments causing the emission has been given to the relevant sanitary authority.[54]

**2.1.6** Pollution control authorities generally require that the discharger monitor his own emissions. Indeed, this would appear to be the only practical course of action as air monitoring facilities available to the control authorities are totally inadequate for effective investigation. The danger of allowing polluters to provide, monitor and report on their own emissions is obvious.

**2.1.7.a** The enforcement of air pollution controls is in the hands of a number of authorities, some of whose functions overlap. The enforcement of planning conditions and of conditions of IDA grants to industry have been described.[55] Failure or refusal to comply with a condition in a licence to smelt ore or ore concentrates may result in the licence being revoked,[56] and/or may render the holder of such licence guilty of an offence and punishable on summary conviction by a fine not exceeding £100, or on conviction on indictment to a fine not exceeding £1000, and in the case of a continuing offence, to a further fine not exceeding £500 for each day during which the offence continues.[57] Failure, neglect or refusal to observe or comply with a condition in a licence to manufacture cement is an offence for which the holder of a licence is liable on summary conviction to a fine not exceeding £100 and, in the case of a continuing offence, to a further fine not exceeding £50 for each day that the offence continues,[58] and the Minister for Industry and Commerce may revoke the licence if the holder has been guilty of such an offence.[59] It is an offence for any person to work, or to attempt to work minerals without lawful authority.[60] It is likewise an offence to search for, or get, petroleum without lawful authority.[61] The aforementioned offences are punishable by fines not exceeding £100 and by

**2.1.7.a continued** a further fine not exceeding £100 for each day that the offence is continued and/or by six months imprisonment. Whenever damage to the surface of land or to mineral deposits or to water supplies or a nuisance is caused, whether directly or indirectly, by the licensee under a prospecting licence;[62] the licencee under a state acquired minerals licence, or the licensee under a mining facilities licence;[63] the leasee of a state mining lease or the holder of a state mining permission;[64] the holder of a petroleum prospecting licence;[65] the leasee of a petroleum lease;[66] the holder of a working facilities permit or the holder of a working facilities (state land) permit;[67] the holder of a mining facilities permit or the holder of a mining facilities (state land) permit in the exercise of conferred rights, or by the Minister when exercising similar rights, compensation shall be payable for such damage or nuisance in accordance with the provisions of the relevant Acts. 'Surface of land' includes 'any buildings, works or any thing erected, constructed or growing on such land.' The rights of those whose property is damaged by air pollutants including dust, etc. would thus seem to be protected. Mineral prospecting licences and petroleum prospecting licences must contain indemnity clauses whereby the licensee indemnifies the Minister against any claim or demand whatsoever in respect of the land, the subject of the licence, or in any way arising out of the exercise by the licensee of the rights conferred on him by the licence.[68] Exploration licences must contain indemnity clauses indemnifying the Minister against claims arising by the exercise by the licensee of his rights under the licence.[69] State mining leases and petroleum leases must both contain such (if any) covenants, conditions and subsidiary agreements as the minister shall consider proper or desirable in the public interest and action may be brought by the minister for breach of same.[70] Both these leases may contain clauses providing for the renewal or successive renewals thereof subject to conditions.[71] The Minister may, at any time, in his absolute discretion, revoke a minerals prospecting licence.[72] He may also revoke an exploration

**2.1.7.a continued** licence, a petroleum prospecting licence and a reserved area licence if he is satisfied that any of the conditions attached to it have been contravened.[73]

Enforcement of the Control of Atmospheric Pollution Regulations is the responsibility of the sanitary authorities who, if they are of the opinion that articles 3 and 4 of the regulations have been contravened and that the contravention is likely to occur again, may notify the occupier of the premises of the works which they consider necessary to be done to end the contravention or to prevent its recurrence.[74] Authorised officers, appointed by the Minister for Local Government or by a sanitary authority, may at all reasonable times enter and inspect any premises for the purpose of ascertaining:

(i) the extent and nature of the contravention,

(i) whether there is or has been, in connection with the premises, any contravention of the regulations,

(ii) the extent and nature of the contravention,

(iii) whether circumstances exist on, or in connection with, the premises which would require any action to be taken under the regulations.

Wilful obstruction or interference with the exercise by an authorised officer of the aforesaid power is an offence punishable on summary conviction by a fine not exceeding £20,[75] and contravention or wilful obstruction of the execution of the regulations is an offence punishable on summary conviction by a fine not exceeding £100 and, if the contravention or obstruction continues after conviction, to a fine not exceeding £5 for each day on which the offence continues.[76] Offences may be prosecuted by the Minister or by the appropriate sanitary authority.[77] A ship or vessel lying in any river, harbour or other water not being within

**2.1.7.a continued** the district of a sanitary authority is deemed to be within the district of the nearest authority.[78] The Minister has suggested to sanitary authorities that proceedings should only be instituted as a 'last resort', and has advised them to seek, rather, the cooperation of industry. Where a breach of the regulations is observed, it was suggested that the authorities issue informal notice of the breach, followed, if necessary, by advance notice of intention to prosecute.[79] Health inspectors attached to sanitary authorities were briefed on the implementation of the regulations.

Sanitary authorities have a statutory duty to cause their districts to be inspected for public nuisances and, on complaint of same and on being satisfied that the complaint is justified, to suppress the nuisance.[80]

The Alkali, etc. Works Regulation Act, 1906, is enforced by central government in the persons of a chief alkali inspector (who is also the chief engineering adviser to the Department of Local Government), and one other inspector who was appointed in 1973. They are responsible for routine inspections to secure compliance with the Act. The alkali inspector has power to require the condensation to his satisfaction of muratic acid gas[81] and of acid gases of sulphur or of sulphur and nitrogen,[82] to serve nuisance abatement notices[83] to inquire whether means can be adopted at reasonable expense for preventing the discharge of noxious or offensive gases from furnaces or chimneys of cement and smelting works, or for rendering such gases harmless and inoffensive,[84] to attach conditions to a certificate of registration or, indeed, to refuse such certificate,[85] to enter (unannounced) and inspect certain works, to examine processes, ascertain the quantity of gas discharged into the atmosphere, and apply such tests and make such experiments and inquiries as he thinks necessary or proper for the execution of his duties.[86] Offences under the Act are punishable by fines which are, by present day standards, minimal.[87] Any sanitary authorities to whom it appears that any work to which the Act applies is carried on in contravention of its provisions, and that a nuisance is thereby occasioned to any of the inhabitants of the district, may complain to the Department of Local Government who must investigate the complaint and may, subsequently, if they think fit, direct proceedings to be taken.[88]

**2.1.7.b** Individuals have powers to object to the decisions of planning authorities.[89] Although the individual is not usually given rights under Act empowering ministers to grant licences, in practice a concerned individual or group of individuals may have access to a minister, who may take account of their suggestions and complaints. The Alkali, etc. Works Regulation Act, 1906, provides that a sanitary authority may complain to the central authority that the Act is being contravened and that a nuisance is being occasioned thereby if ten inhabitants of that sanitary area so complain.[90] The Act expressly gives an injured individual a right to proceed against the owner or owners of a work controlled by the Act where a nuisance is caused by a discharge of any noxious or offensive gases from such work.[91] Apart from the aforementioned, the individual may have a right of action for negligence, nuisance, trespass or under the rules in *Rylands* v. *Fletcher*.[92]

**2.1.7.c** There is no coherent policy on enforcement of air pollution controls. The country lacks the personnel and equipment for detecting breaches of air pollution controls; and laboratory facilities are inadequate. Statistics on the number of prosecutions brought under the Control of Atmospheric Pollution Regulations are unavailable. The Alkali, etc. Act is enforced by two people, one of whom has many other duties to perform. Inspections of three registered works were carried out in 1972 by a firm of engineering consultants and the alkali inspector visited all works at least once in 1973. One complaint against a registered premises was referred to the alkali inspector in the years 1972 and 1973. There have been no prosecutions for contravention of the Act in recent years and, as far as can be determined, a certificate of registration has never been refused. All premises are required to monitor their own emissions.[93] Individuals can rarely afford, in a country where legal aid is not available in respect of civil actions and where the costs of litigations are prohibitive, to sue for damage caused by air pollution.

## 2.2 MOBILE SOURCES

### 2.2.1 Motor vehicles

Motor vehicles are not designed nor, in the strict sense, are they constructed in this country. There are, however, a number of assembly plants where motor vehicles are assembled from parts manufactured in other countries. There are no specific controls over the design of these parts although certain other legal requirements may indirectly operate as controls over design. The Road Traffic Act, 1968, empowers the Minister for Local Government to make regulations to control the importation, supply and fitting of substandard parts and equipment for vehicles,[94] to introduce a 'type-approval' scheme:[95] and to give effect to the European agreement concerning uniform conditions of approval for motor vehicle equipment and parts, 1958.[96] The Minister has never exerised his powers to make the above regulations.

#### 2.2.1.a Controls over design and construction

Legal controls over the construction and equipment of vehicles are contained in Part II of the Road Traffic Acts, 1961 and 1968, and in the Road Traffic (Construction Equipment and Use of Vehicles) Regulations, 1963,[97] as amended. The 1961 Act placed responsibility for compliance with the relevant regulations solely on the user (or owner) of the vehicle, but this responsibility (and controls) were extended by the 1968 Act to ensure that

**2.2.1.a continued** vehicles and vehicle parts were supplied in proper condition.[98] The Road Traffic (Construction Equipment and Use of Vehicles) Regulations, 1963, apply in relation to the use of vehicles in public places,[99] and in so far as they attempt to prevent air pollution provide, in respect of the construction of vehicles: that 'every vehicle shall be so constructed as to prevent, to such an extent as is reasonably possible, the emission of smoke, visible vapour, noxious gases or offensive odours'; that 'every vehicle using solid fuel shall be fitted with an efficient appliance for the purpose of preventing the emission of sparks or grit and also with a tray or shield to prevent ashes or cinders from falling to the ground'; and that a vehicle 'propelled by a compression ignition engine shall not be fitted with a device designed to facilitate the starting of the engine by causing it to be supplied with excess fuel unless:

(a) the device is so designed that it cannot cause the engine to be supplied with excess fuel after the engine has been started, or

(b) the device cannot be readily operated by a person while he is carried on the vehicle.'[100]

## 2.2.1.b Controls over maintenance

The said regulations provide, in respect of the maintenance of vehicles that:

(i) 'Every vehicle and trailer shall be maintained in such a condition that there will not be emitted any smoke, visible vapour, grit, sparks, ashes, cinders, or oil substances, the emission of which could be prevented by the taking of any reasonable steps or the exercise of reasonable care, or the emission of which might cause damage to person or property or endanger the safety or health of any other user of the public place in consequence of any harmful content contained therein;[101]

**2.2.1.b continued** (ii) that devices designed to facilitate starting by causing compression ignition engines to be supplied with excess fuel, be maintained in such a condition as not to cause the engine to be supplied with excess fuel while in motion; [102]

(iii) and that every vehicle used in a public place must be such, and so maintained and used that no danger is likely to be caused to any person.' [103]

## 2.2.1.c Controls over use

It is an offence under the said Regulations to:

(i) use a vehicle in a public place so that there is emitted therefrom any smoke, visual vapour, grit, sparks, ashes, cinders or oily substances the emission of which could be prevented by the exercise of reasonable care;[104], or

(ii) use or cause to be used devices to facilitate the starting of compression ignition engines by causing them to be supplied with excess fuel while the engine is in motion. [105]'

The Minister has power to make regulations prescribing speed limits for all public roads.[106] The overall national speed limit is sixty miles per hour but less speed is permitted in specified areas. The 1968 Act gave the Minister a general power (originally vested in the Commissioner of the Garda Siochana), to make regulations for the regulation and control of traffic in public places, and without prejudice to the generality of the aforesaid power, to make regulations which may, *inter alia,* specify rules for the use of roads by traffic, and prohibit specified traffic from using specified parts of the roadway.[107] Bye-laws by which the Commissioner exercised a similar power were continued in force.[108] The Commissioner still has power, with the consent of the Minister, and after consultation with the local authority concerned to make bye-laws and, with the

2.2.1c continued consent of the Minister only, to make temporary rules, regulating the control of traffic, and to facilitate the movement of traffic in respect of any specified area. Such bye-laws or temporary rules may, *inter alia,* regulate the courses to be taken by traffic on public roads, restrict traffic from entering or passing along any specified road, and prohibit the passage through any public road of any article exceeding a specified length or a specified breadth.[109]

## 2.2.1.d Enforcement process

Enforcement of the Road Traffic Acts, 1961 and 1968, is the responsibility of the Garda Siochana. An offence is committed when a mechanically-propelled vehicle is driven on a public road while there is a defect affecting the vehicle which the owner thereof knows of, or could have discovered by the exercise of ordinary care and which is such that the vehicle, is, when in motion, a danger to the public. The offence is punishable by a fine not exceeding £50 and/or by three months' imprisonment.[110] Breach of the Road Traffic (Construction Equipment and Use of Vehicles) Regulations is an offence punishable by a fine not exceeding £20 in the case of a first offence, and by further penalties for subsequent offences.[111] An authorised officer of the Minister may test a vehicle on the road to ascertain whether, *inter alia,* the regulations are being satisfied.[112] Any Garda Siochana, if he suspects that there is a defect affecting a vehicle which is such that the vehicle when in use is a danger to the public, may inspect and examine the vehicle or require that it be brought to a nearby suitable place for inspection or examination. If the Garda's suspicions are justified, he may require that the vehicle isn't driven in a public place until the defect is remedied. Failure to obey is an offence punishable on conviction by a fine not exceeding £50 and/or imprisonment for up to three months.[113] Regular testing of vehicles is not required except for public service vehicles which must be tested at least once a year.[114] The Commissioner of the Garda Siochana grants licences continued renewable triennially in respect of public service vehicles, and must be satisfied that they comply in all respects with the requirements of the Road Traffic (Construction Equipment and Use of Vehicles) Regulations.[115] He may require that public service vehicles be inspected and examined at any time and as often as he thinks fit,[116] and may refuse to grant or suspend or revoke a licence if the vehicle does not comply with regulations designed, *inter alia,* to prevent air pollution.[117] The owner has a right of appeal against such refusal, revocation or suspension.[118] There are a number of public service vehicle inspectors in the Garda force.

It is extremely difficult to secure a conviction for breach of regulations governing air pollution by vehicles because of the lack of defined standards and the consequent vagueness of the regulations. Control is exercised by visual smoke check only and is exercised on a haphazard basis. The Garda Siochana have not the equipment, manpower or training for detecting breaches, and prosecutions are rare. The regulations themselves are, by modern standards, inadequate and out of date. In any case, effective control of air pollution caused by vehicles must start at the manufacturing stage, and, in this respect, the Irish authorities rely on controls imposed by other governments on the manufacture of vehicles and component parts. There is some justification for this reliance as the vast majority of vehicles and vehicle parts are imported from EEC countries where controls are stringent. Nevertheless, it should not be too difficult for a foreign manufacturer to dump component parts of vehicles obsolete in his own country because of environmental controls on the Irish market. The implementation of EEC measures to control air pollution by vehicles is being considered.

## 2.2.2 Aircraft

There are no direct statutory controls over gaseous emissions from aircraft.

## 2.2.3 Hovercraft

There are no hovercraft in Ireland.

## 2.2.4 Lead in petrol

In answer to a parliamentary question on 21 February 1974, the Minister for Transport and Power gave the impression that the lead content in petrol sold was less than 0.40 grammes per litre.[119]

1. *Irish Journal of Medical Science,* Vol. 140, No. 3, p.117. See also, *Technology Ireland,* May, 1970, pp.10-17 for relevant statistics.
2. See 1-4-2 and 1-6-2.
3. S 10 Local Government (Sanitary Services) Act, 1962. No regulations dealing with this aspect of pollution have been made.
4. Smelting Act, 1968, S 3.
5. Cement Act, 1933, S 11.
6. Id., S 13.
7. Minerals Development Act, 1940, S 8. Such licences are obliged to contain clauses preventing interference with amenities.
8. Id., S 22(1).
9. Id., S 22(3).
10. Id., S 13(1). Such leases must contain such, if any, covenants, conditions and subsidiary agreements as the Minister shall consider proper or desirable in the public interest.
11. Petroleum and Other Minerals Development Act, 1960, SS 75, 76.
12. M.D.A., 1940. S 28(1).
13. Id., S 48(1).
14. Id., S 49(1) and S 6 as amended by S 72 of the P.O.D.A., 1960.
15. Id.,SS 8(1), 24(1), 24(1), 13(2), 28(2), 48(2), 49(2). P.O.D.A., 1960, SS 75(2), 76 (4).
16. P.O.D.A., 1960, S7.
17. Id., S 106 (1).

Id., S 8 (1).

19. Id., S 9(1). Such licence must contain a clause protecting the amenities of the
18. area.
20. Id., S 13(1). Such leases must be granted in the public interest and must contain such (if any) covenants, conditions and subsidiary agreements as the Minister shall consider proper or desirable in the public interest.
21. Id., SS 26(1), 27(2). These are granted to leasees of petroleum leases.
22. Id., S 29. These are ranted to leasees of petroleum leases.
23. Id., SS 8(1), 9(3), 13(1), 13(2)(b), 26(2) and 27(4), 19(2).

24. M.D.A., 2940, S 23.
25. M.O.D.A., 2969, S 52(17.
26. See 1-5-3.
27. Alkali, etc. Works Regulation Act, 1906, S 9'
28. Id., S 27.
29. Id., First Schedule. Some twenty one works are listed as scheduled works.
30. Id., S 12.
31. Report for the years 1972 and 1973 by the Chief Alkali Inspector.
32. See e.g. Circular to Sanitary Authorities 2-2-71, L4/71 para. 15 and Report of the Chief Alkali Inspector for 1972 and 1973.
33. See 1-2-3.
34. SS 2, 7.
35. S 8.
36. See 2-1-1.
37. ENV: 17/73, 11 June 1973.
38. S 1. No. 156/1970.
39. Defined in Article 1.
40. Article 7.
41. S 1(1)
42. S 6(1)
43. S 6(2)
44. S 7(1)(b).
45. S 2(1).
46. S 7(a).
47. HMSO 1967.
48. Circular on Chimney Heights, ENV. 17/73, 11 June 1973.
49. S 1, No. 156/1970.
50. Id., Article 3(3)(a).
51 Id., Article 3(3)(b).
52. Id., Article 3(3)(c).
53. Id., Articles 3(4), 3(5).
54. Id., Article 7.
55. See 1-4-2 and 1-5-3.
56. Smelter Act, 1968, S 5.
57. Id., S 6.
58. Cement Act, 1933, S 14.
59. Id., S 16(2).
60. Minerals Development Act, 1940, S 74.
61. Petroleum and Other Minerals Development Act, 1960, S 6.
62. M.D.A., 1940, S 10(1) as amended.
63. Id., S 25(1) as amended.
64. Id., S 31(3) as amended.
65. P.O.D.A., 1960, S 12.
66. Id., S 16.
67. Id., S 28(1).
68. S 8(4) and S 9(6) of the respective Acts.
69. P.O.D.A., S 8(6).
70. S 26 (1)(b) and S13 (1) (b) of the respective Acts.
71. S 26(1)(c) and S 13(1)(c) of the respective Acts.

72. M.D.A. 1940, S 12(1).
73. P.O.D.A., S 22.
74. Control of Atmospheric Pollution, Article 6, Regs. 1970.
75. Local Government (Sanitary Services) Act, 1962, S 10(5).
76. Id., S 10(6)
77. Id., S 10(7).
78. Id., S 10(9).
79. Circular to Sanitary Authorities, L4/71 (2/2/71).
80. Public Health (Ireland) Act, 1878, S 108.
81. Alkali, etc. Works Regulation Act, 1906, S 1(1).
82. Id., S 6(1).
83. Id., S 5.
84. Id., S 8(1).
85. Id., S 9(5).
86. Id., S 12.
87. Id., SS 1(2), 2(2), 3(2), 4(2), 6(3), 7(2).
88. Id., S 22(1).
89. See 1-6-2.
90. See 2-1-7(a) *supra*.
91. Alkali, etc. Act, 1906, S 23(1).
92. See 1-6-2.
93. Information from the Alkali Inspector and the Annual Report for the years 1972 and 1973.
94. S 9.
95. S 8
96. S 14.
97. SI, No. 190 of 1963.
98. SS 8, 9, 14, 17.
99. Public place means any street, road or other place to which the public have access with vehicles, whether as of right or by permission and whether subject to, or free of, charge.
100 Article 30
101 Article 34 (f).
102. Article 34 (g)
103. Article 90 (I)
104. Article 90(1).
105. Article 90(3).
106. R.T. Act, 1961, SS 44, 46. R.T. Act, 1968, SS 24, 25.
107. R.T. Act, 1968, S 60.
108. Id., S 60(4) and Road Traffic (Bye-laws and Temporary Rules) Regulations, 1961, No. 219 of 1961 and Road Traffic General Bye-laws, 1964, SI No. 294 of 1964
109. R.T. Act, 1961, S 89.
110. R.T. Act, 1961, S 54 as amended by Schedule to R.T. Act, 1968.
111. R.T. Act, 1961, SS 11(4), 103, R.T. Act, 1968, S8.
112 R.T. Act, 1968, S 18.
113. R.T. Act, 1961, S 20.
114. Road Traffic (Public Service Vehicles) Regulations, 1963, SI No. 191 of 1963, Article 33.

115. Id., Article 20.
116. Id., Article 33.
117. Id., Articles 20,24,25.
118. Id., Article 27.
119. Dail Reports, Vol. 270, Cols. 1433-1434.

# 3 Inland waters

## Explanatory Note

The principal causes of water pollution in Ireland are stated to be,[1] in approximate order of importance:

a) industrial wastes,

b) discharge of untreated sewage,

c) farm wastes,

d) inadequate or improper treatment of sewage,

e) effluent from mining and milled peat industries and

f) dumping of rubbish.

In 1971 An Foras Forbartha surveyed the main water systems in the country — one hundred and twenty-one rivers to a total length of 2900 kilometres—and reported that about seven percent of the total length of rivers surveyed were seriously polluted, that about ten percent gave cause for concern and that the remainder surveyed were free from significant pollution.[2] These statistics are more ominous than is apparent as most of the serious pollution occurs in the densely populated areas. Industrial wastes when quantified by the population equivalent of effluent are greater than the figure of 1.68 people.[3] The sewage systems of almost all towns serving populations of more than 10,000 are causing or contributing to pollution.[4] And of the one hundred and twenty-eight towns with populations ranging from 1000 to 10,000 surveyed, eighty-four, broadly speaking have deficient sewage systems.[5] Pollution from creameries is almost as great as that from the sewage of all inland towns, and only in a few cases have proper effluent treatment facilities been provided.[6] Problems also exist with regards to effluents from meat and food processing factories.[7] Pollution from agricultural wastes has not been quantified but, as much agricultural activity is free from legal control, damage must be extensive since Ireland was producing 1100, 24,000, and 8400 million gallons, respectively, of pig, cattle and poultry slurry, in 1972,[8] and 4.5 million tons of silage effluent in 1971.[9]

Various statutes provide sanctions for causing pollution, but all contain very vague terminology such as 'noxious,' 'poisonous' or 'deleterious' without any attempt to give a scientific definition of those terms. Only two statutes[10] are extensively used to prevent water pollution, but widely framed clauses in others may provide additional controls. The Interdepartmental Committee on Water Pollution, which consisted of civil servants, concluded that 'generally speaking, the law relating to water pollution is not satisfactory. Though common law actions can sometimes be effective and the statutory provisions on the subject are numerous and detailed, their combined effect has not sufficed to curb the gradual spread of water pollution or to offer a reasonable hope of controlling the continuance of this trend... This failure is due primarily to the rigidity and consequent difficulty in enforcement of the statutes, and also to the cumbersome procedures provided for in the legislation.'[11]

It must, however, be recorded that there has been, and is, a great deal of concern about water pollution in recent years, evidenced by the many investigations and studies which have been undertaken to establish the extent of the problem. Government departments have not been inactive. The Department of Local Government has had surveys carried out on, *inter alia,* water quality in the national rivers, local authority sewage disposal, and on the extent to which industry is contributing to pollution. The Department of Agriculture and Fisheries has investigated pollution caused by agriculture, especially pollution caused to fisheries. Working groups set up by the Government have produced the aforementioned report on water pollution and a report on oil pollution.[12] The nature and extent of the problem has now been determined — action is under active consideration.

**3.1.1** Control over the siting of points of discharge of pollutants into water may be exercised by conditions attached to planning permissions or approvals,[13] and by recommendations in development plans; by conditions attached to smelting licences, cement licences, and to various licences, leases, permits and permissions which may be granted under the Petroleum and Other Minerals Development Acts, 1940-1960;[14] by sanitary authorities when permitting communication with sewers belonging to them;[15] by the IDA when grant-aiding industries; by the Minister for Agriculture and Fisheries when granting licences under the Fisheries Acts, 1959 and 1962, and by the agricultural advisory services when approving farm buildings for grant eligibility.

**3.1.2.a** Control over processes may be exercised by all the methods described in the preceding paragraph. In addition, criminal sanctions for causing pollution may necessitate the taking of preventative measures.

The Planning Act, 1963, specifically empowers planning authorities to include in their development plans objectives for the purposes of 'prohibiting, regulating or controlling the deposit or disposal of waste materials and refuse, the disposal of sewage and the pollution of rivers, lakes, ponds, gullies and the seashore.'[16] Control over individual processes may be exercised by conditions attached to planning permissions and approvals. A procedure has been evolved whereby industrialists seeking grant assistance from the IDA are required to give information and undertakings in regard to effluent

**3.1.2.a continued** discharge in a prescribed form, then An Foras Forbartha and the IIRS in consultation with the Department of Agriculture and Fisheries or other interests as appropriate prepare a joint recommendation on the standard to which effluent should be treated, having regard to all relevant factors, including the condition of the receiving waters. The IDA then sends the completed forms to the appropriate planning authority. The planning authority, when giving notice of a decision to grant a permission or approval for industrial development liable to give rise to effluent problems sends copies of the notification to the Fisheries Division and to An Foras Taluntais. [17]

The Alkali, etc. Works Regulation Act, 1906, S 3, makes it an offence punishable by fine to discharge chemical waste from a 'scheduled industry,' [18] whether into water or otherwise; 'so as to cause a nuisance.' The owner may require a sanitary authority to provide (at the former's expense), a suitable drain channel discharged into the sea or river or watercourse, in order to reduce pollution. This discharge, however, must not be contrary to the Rivers Pollution Prevention Act, 1876, as amended.

The Public Health (Ireland) Act, 1878, S 77, makes it an offence for any person engaged in the manufacture of gas to cause or suffer to be brought or to flow into any stream, reservoir, aqueduct, pond or place for water or into any drain or pipe communicating therewith, any washing or other substance produced in making or supplying gas, or wilfully to do any act connected with the making or supply of gas whereby water is fouled. The penalty is £200 plus £20 daily for a continuing offence. Similar provisions are contained both in S 62 of the Waterworks Clauses Act, 1847, and in S 21 of the Gasworks Clauses Act, 1847, which are incorporated in the Public Health (Ireland) Act, 1878, and the Gas and Water Facilities Act, 1870, respectively. The Public Health Acts Amendment Act, 1890, S 17, makes it an offence for any person to turn or permit to enter any

**3.1.2.a continued** sewer of a local authority or any drain communicating therewith any chemical refuse or any waste steam, condensing water, heated water or other liquid (being of a higher temperature than 100°F.) which, either alone or in combination with the sewage, causes a nuisance or is dangerous or injurious to health.

The Rivers Pollution Prevention Acts, 1876 and 1893, prohibit the discharge of the solid refuse of any manufactory, manufacturing process or quarry, or any rubbish or cinders or any other waste or any putrid solid matter into any stream so as to interfere with its flow or pollute its waters,[19] or to discharge into any stream any poisonous, noxious or polluting liquid proceeding from any factory or manufacturing process,[20] or to discharge from any mine into any stream solid matter so as to interfere with its due flow, or any poisonous, noxious or polluting solid or liquid matter other than water in the condition as it was in the mine.[21] It is a defence in the last two instances to prove that the best practicable and reasonably available means were used to render the discharge harmless. A certificate of an inspector from the Department of Local Government is made conclusive evidence of 'the best practicable and reasonably available means' in all proceedings under the Act.[22] Stream includes rivers, streams, canals, lakes, watercourses (other than a watercourse used before 1876 mainly as a sewer and emptying directly into the sea), and may include so much of the sea and tidal waters as the Minister for Local Government may determine by order.[22] No such order has been made. Proceedings for offences under the Act must be taken by the appropriate sanitary authority, but the Act provides a number of restrictions on proceedings which unnecessarily favour industries causing pollution.[24] Sanitary authorities have a statutory duty to afford facilities for factories draining into sewers unless any liquid would prejudicially affect the sewers or the disposal of sewage, or would, from its temperature or otherwise, be injurious from a sanitary point of view, or would overload

**3.1.2.a continued** sewers or interfere with any court order respecting the sewage of such authority.[25] Courts having jurisdiction to try offences under the Act have powers to order the abstention from offences, performance of duties, to impose conditions as to the time or mode of action, to rescind or suspend any order, and, generally, to give such directions as are meet for the carrying into effect of any order. A court may also get an expert's report on the best practicable and available means for preventing pollution, but such report must take into account the reasonableness of the expense involved in providing such means. Default in complying with a court order renders the defaulter liable to pay up to £50 per day for each day in default. If the default is persistent, the court may appoint a person to execute the court order at the defaulter's expense.[26] Provision is made for appeals by aggrieved parties.[27] The Acts are, as the interdepartmental committee mildly concluded,[28] 'basically defective'. Terminology used is imprecise and no scientific definition of 'deleterious', 'noxious', etc. is provided. No definition is given to the 'pollution' except the provision that 'polluting' is not to include 'innocuous discoloration'.[29] The enforcement provisions are administratively very complex with the result that 'it is not surprising that sanitary authorities should be reluctant to commence proceedings by so complex a method.[30]'

The Electricity (Supply) (Amendment) Act, 1945, S 42, prohibits any person from discharging or allowing to escape into a river or stream serving electricity generating stations any chemical or other substance likely to injure part of the generating station or any subsidiary or connected works without the written permission of the Electricity Supply Board and compliance with conditions attached to such permission. Contravention entails a £50 fine, plus £20 for each day on which the offence is continued.[31]

The Harbours Act, 1946, S 88, makes it an offence punishable by a fine, up to £10, for any person to put

3.1.2.a continued ballast, earth, ashes, stones or any other substance or thing within the limits of a harbour without the authorisation of harbour authority, and S 60 empowers harbour authorities to make bye-laws for regulating, *inter alia,* the ballasting of vessels within the harbour and the order and manner in which vessels shall be supplied with ballast and the discharging, removal and disposal of ballast.[32] Proceedings for contravention of the Act may be taken by the Minister for Transport and Power, or where the contravention occurs within a harbour, by the appropriate harbour authority.[33]

The Fisheries (Consolidation) Act, 1959, as amended by the Fisheries (Amendment) Act, 1962, contain two provisions dealing with water pollution. Section 171 of the 1959 Act prohibits the steeping of flax or hemp or the discharge into any waters of any deleterious matter otherwise than under and in accordance with a licence granted by the Minister for Agriculture and Fisheries. In 1973 eighteen such licences were granted — eleven to industries and the remaining seven to local authorites in respect of sewage systems. The said Minister is, after consultation with the Minister for Industry and Commerce in respect of applications for licences from industries, and after consultation with the Minister for Local Government in respect of applications from local authorities, empowered to grant the licence, attach conditions, and, if necessary, and after consultation with the other appropriate Minister to revoke a licence. At no time during the last fourteen years has a licence been revoked. Conditions are, in practice, laid down in each case depending on the water concerned. These stipulate the quantity of effluent that may be discharged in twenty-four hours, and contain qualitative stipulations as to the content of any discharge. A Department spokesman has stated that 'the object of these licences is to allow people to discharge an effluent in a particular place which would be deleterious by definition but would not do any harm to the water concerned. If, for

**3.1.2.a continued**

instance, a poisonous substance even in very small quantities was contained in the effluent, we would not grant a licence.'[34] Section 172 of the 1959 Act, as amended, deals with deleterious matter which is contained or conveyed in a receptacle which is within thirty yards of any waters. The owner of the receptacle must prevent the passage of the deleterious matter into the waters and the Minister may grant a certificate as to the 'suitable means' for preventing pollution of the waters from such receptacle. No such certificate has been issued. The word 'waters' means any river, lake, watercourse, estuary or any part of the sea.[35] Deleterious matter is defined as 'any substance (including explosive, liquid or gas) the entry or discharge of which into any waters is liable to render those or any other waters poisonous or injurious to fish, spawning grounds or the food of any fish or to injure fish in their value as human food, or to impair the usefulness of the bed and soil of any waters as spawning grounds or their capacity to produce the food of fish.'[36] Legal proceedings for offences under the Fisheries Act, 1959 and 1962, may be brought by the Garda Siochana, the Boards of Fishery Conservators and the Minister for Agriculture and Fisheries.[37] The penalties for breach of Sections 171 and 172 are, respectively, a fine up to £100 and/or six months imprisonment, and a fine up to £50 plus £10 per day for a continuing offence, but subject to a total maximum fine of £100. Information on the number of prosecutions taken is not readily available, but the Committee on Water Pollution concluded that the powers under the Fisheries Acts for dealing with pollution were 'reasonably well enforced' by the Boards of Fishery Conservators.[38] It is obvious that many processes which ought to apply for licences under S 171 have not done so.

## 3.1.2.b Sewage

Control over pollution caused by sewage is generally the responsibility of sanitary authorities.[39] The principal Acts governing the powers, duties, and responsibilities of these authorities are still the Public Health (Ireland) Act, 1878,

**3.1.2.b continued** and the Public Health Acts Amendment Act, 1890. Sewers within its district, with minor exceptions, together with structures ancillary thereto are vested in, and under the control of, each sanitary authority,[40] which has wide powers to carry out its functions in providing sewage systems, including powers to purchase and make sewers,[41] to alter and discontinue sewers,[42] to dispose of sewage,[43] to agree to the communication of sewers with adjoining districts,[44] to deal with land appropriated to sewage purposes,[45] to enter into contracts to carry out its duties,[46] and to enforce the provisions of the Public Health Acts, 1878 and 1890, relating to nuisances.[47] Sanitary authorities are empowered to make bye-laws with respect to many matters including drainage of buildings, waterclosets, earthclosets, privies, ashpits, and cesspools connected with buildings, and the keeping of water closets supplied with sufficient water for flushing.[48] They, themselves are obliged:

(i) to keep all sewers in repair and to cause to be made such sewers as may be necessary for effectually draining their district,[49] (this section is mandatory and a procedure is provided for compelling sanitary authorities to discharge this duty[50]);

(ii) to purify sewage before discharging it;[51]

(iii) to cause their sewers to be so constructed, covered, ventilated and kept so as not to be a nuisance or injurious to health, and to be properly cleansed and emptied.[52]

Most sections provide, and in any case it is a general principle of the common law, that statutory powers and duties can not be used so as to occasion a nuisance, unless such nuisance is the necessary result of works authorised by statute. Once a sewer exists, any individual has a right to empty his drains into it by giving notice to the sanitary authority and on complying with the regulations of the

**3.1.2.b continued** authority as to the manner in which the communication is to be effected.[53] An individual may require the authority to effect the communication for him on paying the costs thereof.[54] The sanitary authority has power, where a public sewage system has been provided, and where a connection of less than one hundred feet would suffice to connect unsatisfactorily drained premises with the public sewer, to serve a notice on the owner of the premises requiring him to have his premises serviced by the public drain.[55] It is an offence, punishable by a fine of £10 plus 50p for each day the offence continues, to cause to pass into a local authority sewer any matter or substance by which the free flow of sewage or surface or storm waters may be interfered with, or by which any sewer or drain may be injured.[56] Sanitary authorities may be controlled by virtue of S 171 of the Fisheries (Consolidation) Act, 1959, but only seven authorities are so controlled.[57]

**Water supplies**

Control over water supplies is generally the responsibility of sanitary authorities in whom reservoirs and works used for supplying water to the public are vested.[58] Authorities have a general power to supply their districts with 'a supply of water proper and sufficient for public and private purposes'.[59] Where a public water supply system has been provided, sanitary authorities may in certain circumstances require the connection with such system of any premises not provided with a satisfactory water supply.[60] They are specifically obliged to provide and keep in waterworks belonging to them a supply of 'pure and wholesome water'.[61] They may take proceedings to prevent pollution of waterworks within their jurisdictions from sewage,[62] and may, on receipt of a complaint that potable water is so polluted as to be injurious to health, seek a court order to eliminate such danger.[63] Where a sanitary authority is empowered to take water from any source, it has the same rights of preventing interference with the flow and pollution of the water as has the riparian owner.[64] Section 61 of the

**3.1.2.b continued** Waterworks Clauses Act, 1847, prohibits the contamination of any stream or reservoir used as a public water supply, or any aqueduct or other part of the supply system. Section 17 of the Waterworks Clauses Act, 1863, makes it an offence for any person to cause or suffer any pipe, valve, cistern or other apparatus to be out of repair or to be so used as to cause water contamination. Section 16 of the same Act allows a sanitary authority to cut off the water supply to anyone who fails in certain circumstances to prevent contamination of the supply.[65] Sanitary authorities ensure the purity of waters by chlorination which has become accepted as a standard procedure in the treatment of potable waters in municipal supplies.

**Agriculture**

Water pollution caused by agriculture is virtually uncontrolled by law. The use of land for agricultural purposes and the provision of agricultural buildings are not generally subject to control under the Planning Act, 1963.[66] Such legal controls as do exist are punitive rather than preventative and consist of sanctions provided by statute law for water pollution and of common law remedies. A great deal, therefore, depends on measures designed to secure the voluntary cooperation of farmers to remedy existing problems and to prevent future pollution. Grants are available to farmers in respect of farm buildings, silos etc. from the Department of Agriculture. The approval of grant aid for works which are likely to cause pollution is subject to the following conditions:

(i) The buildings being sited where wastes are clearly unlikely to cause pollution or where wastes can be adequately and safely trapped and contained in storage facilities;

(ii) Adequate waste storage facilities being provided for service of buildings concerned and of all other similar structures on the site (the norm is twenty cubic feet of storage per pig fattening place or equivalent);

**3.1.2.b continued** (iii) The applicant owing, or having at his disposal, enough suitable land to spread wastes from the grant-aided projects, without incurring a serious risk of water contamination by run-off percolation, or build up of excess nutrients in the long term. (A minimum of seventy acres of land is required per thousand fattening places or equivalent).[67] County Committees of Agriculture provide advice to farmers on how to control pollution. The Department of Agriculture and Fisheries has produced booklets dealing with silage effluent and farmyard waste problems, which are supplied free to farmers, local authorities and interested bodies. The Department of Local Government had circularised local authorities and fishery authorities on how to deal with farmers causing pollution. A ministerial circular[68] to local authorities advised them, on finding that a farmer is causing water pollution, to write to him,

(a) indicating the location, extent and cause of pollution;

(b) requiring him to remedy the situation and to notify them, within a specified time, on how he proposes to do this;

(c) warning him, that unless the pollution is remedied, legal action may be contemplated;

(d) indicating that the advice of his agricultural instructor is available to him by application for same to his local County Committee of Agriculture.

A copy of the notice is to be sent to the chief agricultural officer. Since the statutory basis for operation of the Agricultural Advisory Service is in the functional area of each county council only (i.e. exclusive of urban districts), water pollution caused by agricultural activities in urban areas, is a question to be discussed and agreed locally

**3.1.2.b** **(continued)** with the appropriate chief agricultural adviser.[69] Where a farmer does not act as requested to stop the pollution, it is open to the local authority to take all necessary steps to secure its abatement. The approach to pollution caused by agriculture is, therefore, a voluntary one backed by the Agricultural Advisory Service and grant schemes. This is obviously unsatisfactory in modern conditions. Agricultural activities cause, and known to cause, considerably pollution to waters, yet in 1967, 1968 and 1970 only thirteen, thirteen and three complaints, respectively, in respect of water pollution from effluent were brought to the notice of the Department of Agriculture and Fisheries.[70]

**Offensive trades**

It is an offence under S 128 of the Public Health (Ireland) Act, 1878 to establish within the district of any urban authority, without written consent, any offensive trade. The following trades are offensive trades for the purposes of the Act — the trade of blood-boiler, bone-boiler, fellmonger, soap—boiler, tallow-melter, tripe-boiler, gut manufacturer and any other noxious or offensive trade, business or manufacture *ejusdem generis* with those specified. The penalty is £50 plus £2 for each day on which the offence is continued. Urban authorities are empowered to declare a business an offensive business[71] and may make bye-laws with regard to any trade which is an offensive trade in order to prevent or diminish any noxious or injurious effects therefrom.[72] Where an urban authority medical officer, or two medical practitioners or ten inhabitants of an urban district certify that an offensive trade is a nuisance or injurious to health, an urban authority must bring proceedings against that trade. It is a defence for the offender to prove that he has used the best practicable and available means for abating such nuisance or for preventing or counteracting any effluvia which is a nuisance or injurious to health. The maximum penalty on conviction is £5 in the first instance, but further penalties up to £200 may be imposed on subsequent convictions.[73]

**3.1.3** No statute positively requires the pre-treatment of any discharge into water, but such pre-treatment may be necessary

(i) in order to comply with operating conditions in licences, leases, permits or permissions granted by water pollution control authorities,[74]

(ii) in order to avoid attracting criminal sanctions provided by various legislative enactments herein described for pollution, and

(iii) in order to forestall civil actions available to those who suffer damage from pollution.[75]

**General**
Apart from legislation designed to prevent pollution by specific undertakings, three statutory provisions make water pollution in general, an offence:

(i) The Rivers Pollution Prevention Act, 1876,S 2, makes it an offence for any person to discharge rubbish, cinders or any waste or putrid solid matter into any stream so as to interfere with its flow or pollute its waters.

(ii) The Public Health Acts Amendment Act, 1890, S 47, prohibits the throwing of cinders, ashes, bricks, stones, rubbish, dust, filth or other matter likely to cause annoyance into any river, stream or water course.

(iii) The Rivers Pollution Prevention Act, 1876, S 3,makes it an offence for any person to discharge any solid or liquid sewage into watercourses. Person includes any body of persons whether corporate or incorporate,[76] and, therefore, includes sanitary authorities. This prohibition is unworkable because as the Committee on Water Pollution pointed out 'it is impracticable to produce a sewage effluent free from solid of liquid sewage matter.' The penalties for all these offences are minimal.

**3.1.4** Limits on the quality and quantity of discharges to water may be imposed by the stated control authorities. Mandatory standards are not generally provided by legislation. This is understandable as there is a general reluctance to impose inflexible standards which may require frequent variation. The Industrial Development Authority usually sets standards recommended by the Institute of Industrial Research and Standards and/or An Foras Taluntais.[77] The Minister for Local Government, in an effort to ensure a consistent approach by all planning authorities in considering standards, has specifically referred to them[78] to three documents indicating suitable standards in particular cases, viz. the IIRS booklet *Recommendations for the Disposal of Industrial Effluents,*[79] a British publication *Standards of Effluents to Rivers with Particular Reference to Industrial Effluents;*[80] and a publication called *Agricultural Wastes in Ireland.*[81] Planning authorities normally specify precise requirements governing the disposal of effluent, and may require applicants to furnish information to assist them in setting these requirements. When permission is granted without the aforementioned information being supplied, authorities have been recommended[82] to include a condition requiring developers before commencing development:

(a) to formally agree as to the limits in quality and quantity of industrial effluent which will be discharged into the public sewer,

(b) to make necessary arrangements for regular monitoring of the discharges and

(c) where appropriate, to supply details of proposed methods for treating and discharging effluents.

**3.1.4 (continued)** Authorities were also advised to take prompt action where approved plans are not followed or conditions violated. This practice of granting permissions before all the implications of a proposed development can be appreciated is dangerous and undesirable.

The Rivers Pollution Prevention Act, S 7, permits sanitary authorities to refuse facilities for factories discharging liquids into sewers where such liquids 'would prejudicially affect such sewers or the disposal by sale, application to land, or otherwise, of the sewage matter conveyed along such sewers, or which would, from its temperature of otherwise, be injurious in a sanitary point of view.' The discharge of such liquids may, however, be permitted if dischargers comply with standards required by sanitary authorities. The Public Health (Ireland) Act, 1878, S 19, requires sanitary authorities to purify sewage before discharging it into waters. This section is widely flouted.[83] The Public Health Acts Amendment Act, 1890, S 17, prohibits the discharge into sewers of waste steam, condensing water, heated water, or other liquid of a higher temperature than 110°F., which, either alone or in combination with the sewage, causes a nuisance or is injurious to health.

**3.1.5** Control over the methods by which pollutants are discharged into water is usually exercised by virtue of general powers to control water already described. The local of industries is often of crucial importance when considering methods of discharge and it is relevant to note that a great number of industries liable to have effluent problems are situated, and encouraged to set up in areas where the volume of receiving waters is sufficient to obviate or reduce pollution.

**3.1.6** Planning authorities and the Minister for Agriculture and Fisheries may, and often do, require that the polluter monitor his own discharges. There are, however, no developed administrative arrangements for monitoring water pollution or effluent discharges. Many different control authorities are involved and local authorities, who are in fact the main controlling bodies, lack the necessary powers, resources and determination to institute an efficient and comprehensive monitoring system.

**3.1.7** The general position with regard to the enforcement of water pollution controls has already been described. It is scarcely necessary to record that the multiplicity of controlling authorities is hardly conductive to effective enforcement; neither is the fact that local authorities who, as planning and sanitary authorities, are the chief controlling bodies, are themselves major polluters. The situation, as it exists, often results in a local authority being in a position to sue itself or another local authority — hence no enforcement. The Committee on Water Pollution has recommended that local responsibility for abatement and prevention of water pollution, including control of discharges and abstractions and sampling of receiving waters, should be placed on county councils and county borough corporations,[84] and that overall responsibility at central level be vested in the Minister for Local Government.[85] The wisdom of this recommendation, in the light of the existing circumstances of local authorities and their past record is doubtful.

**3.1.8** Individual rights to control water pollution are numerous and generally far more effective than the rights of controlling authorities.

**3.1.8.a** There is no specific legal requirement giving the citizen a right to be informed of potential water pollution. However the Local Government Planning and Development Act, 1963, (Permission) Regulations, 1964, and the legal obligation on planning authorities to display development plans afford the interested citizen an opportunity of obtaining information and of taking action to prevent water pollution.[86]

**3.1.8.b** The rights to object to planning permissions and to make representations with regard to development plans have already been described.[87]

**3.1.8.c** Individual rights to restrain water pollution exist at common law and by virtue of various statutes.

At common law the individual may have the right to bring an action in trespass, negligence or nuisance or under the rule in *Rylands* v. *Fletcher*.[88] In addition, every riparian owner and every owner of a fishery has a right to have a stream come to him in its natural state in flow, quantity and quality. If the water coming to him is polluted, he may seek an injunction to restrain such pollution. In *Crossley* v. *Lightowler,*[89] which was an action between riparian owners relating to pollution of a stream by dye-works, it was held that a riparian proprietor can maintain a suit to restrain the fouling of water without showing that such fouling is actually injurious to him. Damages may also be recoverable if the riparian owner has suffered material loss, e.g. injury to a fishery or enjoyment of fishery rights.

Certain statutes expressly confer a right to bring an action to prevent water pollution on the individual. Individual rights under the Public Health (Ireland) Act, 1878, have been described.[90] The Rivers Pollution Prevention Act, 1876, S 6, provides that the individual may complain to a sanitary authority that an offence under that Act has been committed, and, on the failure of sanitary authority to take proceedings in respect of such offence, to appeal to the Minister for Local Government who may direct the sanitary authority to take the necessary proceedings. It is generally true to say that receipt by any control authority of a complaint alleging water pollution

3.1.8.c (continued) results in the complaint being investigated, and, on fact being found justifying such complaint, in remedial action being demanded.

**Miscellaneous**

A number of provisions in statutes enacted to deal with specific areas or undertakings are relevant to water pollution control. Those dealing with particular areas are: the Shannon Electricity Act, 1925,[91] the Liffey Reservoir Act, 1936,[92] and the Foyle Fisheries Act, 1952. Those dealing with specific undertakings are the Electricity Supply (Amendment) Act, 1945, [93] the Arterial Drainage Act, 1945,[94] the Turf Development Act, 1946,[95] and the Local Authorities (Works) Act, 1949,[96] which exempt these operations from the provisions of the Fisheries Acts.

## 3.2 CONTROLS OVER DUMPING

See preceding paragraphs in this section and section 6.

## 3.3 WATERBORNE CRAFT[97]

### 3.3.1 Controls over design and construction

Controls over the design and construction of ships may be exercised by the Minister for Transport and Power who is empowered to make rules prescribing the requirements with which the hull, equipment and machinery of passenger ships[98] and cargo ships[99], registered in the state, must comply. The standards of structural strength of ships, including tankers, are governed by International Load Line Convention, 1930, which was adopted in Ireland by the Merchant Shipping (Safety and Load Line Conventions) Act, 1933. The 1930 convention has been modified and extended to cover vessels of larger sizes now being built by the international convention on load lines, 1966, and the Merchant Shipping (Load Lines) Act, 1968, which adopts the latter convention. The Minister has made load line rules[100] to bring into effect detailed provisions of the 1966 convention.

## 3.3.2 Controls over maintenance

Controls over maintenance of ships are in accordance with international law and practice. Ships are required to have safety certificates by the Merchant Shipping (Safety Convention) Act, 1952, and to have load line certificates by the Merchant Shipping (Load Lines) Act, 1968.

## 3.3.3 Controls over use

Ships are (with certain exceptions) prohibited from proceeding to sea without safety or load line certificates.[101] The master of a ship is required to report on dangers to navigation by the Merchant Shipping (Distress and Urgency Signals and Danger Warning) Rules, 1936,[102] and by section 35 of the Merchant Shipping (Safety Convention) Act, 1952, as extended by section 17 of the Merchant Shipping Act, 1966. Where ice is reported on or near the course of a ship, the master thereof is obliged at night to either proceed at a moderate speed or to change his course so as to avoid danger.[103] Owners of passenger steamers, crossing the North Atlantic to or from any Irish port by regular routes, must give public notice of proposed routes or of changes of such routes.[104]

The Minister has made rules[105] regulating the carriage of dangerous goods by:

(a) ships registered in the state, and

(b) other ships while they are within any port in the state, or are embarking or disembarking passengers within the territorial waters of the state, or are loading or discharging cargo or fuel within those waters.[106]

**3.3.4** The Merchant Shipping Acts are enforceable by the Minister for Transport and Power and by inspectors, ship surveyors and engineering surveyors.

1. Report on Water Pollution, Prl. 2939, para. 4-1-1.
2. *The National Survey of Irish Rivers—A Report on Water Quality,* 1972.
3. Report on Water Pollution, *supra* para. 4-2-2.
4. Id., para. 5.4.2.
5. Id. see Appendix 9.
6. Id., paras. 4.2.4 to 4.2.5.
7. Id., paras. 4.2.6 to 4.2.9.
8. *Technology Ireland,* Jan., 1973, p.29.
9. Report on Water Pollution, para. 4-4-5.
10. The Local Government (Planning and Development) Act, 1963, and the Fisheries (Consolidation) Act, 1959, as amended.
11. At para. 6-20-1.
12. Prl. 1431.
13. See 2.1.1 and 1.4.2.
14. See 2.1.1.
15. See 3.2.3.
16. S 19 and Third Schedule, Part IV, Clause 11.
17. See Circulars L16/71, 25/5/71, as amended by ENV. 23/73, 27/8/73, sent to all planning authorities.
18. See 2.1.2.
19. R.P.P.A. 1876, S 2.
20. Id., S 4.
21. Id., S 5.
22. Id., S 12.
23. Id., S 20.
24. Id., SS 6, 13.
25. S 7.
26. S 10.
27. S 11.
28. Report on Water Pollution, para. 6-3-3.
29. R.P.P. Act, 1876, S 20.
30. Report on Water Pollution, para. 6-3-4.
31. As to electricity supply see 1.6.1 (iii)
32. Second Schedule, Clause 40. A typical example of such bye-laws is the Cork Harbour Commissioners bye-laws relating to navigations, etc. of 11/6/56, bye-laws 99-101.
33. See also sections 4 and 5 below.
34. *Sunday Press,* 30/12/73.
35. S 3(1).
36. F. (A) Act, 1962, S 2(1).
37. F.A., 1959, SS 309, 312.
38. Report on Water Pollution, para. 7-15-2.
39. See 1.4.4.
40. P.H. (2) Act, 1878, S 15.

41. Id., SS 16, 18.
42. Id., S 20.
43. Id., S 30.
44. Id., S 31.
45. Id., S 32.
46. Id., S 200.
47. Id., SS 107-127.
48. Id., S 41 as extended by P.H.A., 1890, S 23.
49. Id., S 17.
50. Id., S 211.
51. Id., S 19. This section does not apply to discharges into the sea or tidal waters. The R.P.P. Act, 1876, S 19, gives statutory immunity to sanitary authorities in this case.
52. Id., S 21.
53. P. H. (Ir.) Act, 1878, S 23.
54. P.H.A.A. Act, 1890, S 18.
55. Local Government (Sanitary Services) Act, 1962, S 8(1).
56. P.H.A.A. Act, 1890, S 16(1).
57. See 3.1.2.a.
58. P.H. (Ir.) Act, 1878, S 74.
59. Id., S 61.
60. Local Government (Sanitary Services) Act, 1962, S 8(2).
61. P.H. (Ir.) Act, 1878, S 65.
62. Id., S 78.
63. Id., S 79.
64. Water Supplies Act, 1843, S 20.
65. The said sections of the Waterworks Clauses Acts were incorporated in the P.H. (Ir.) Act, 1878, S 67.
66. S 4 and L.G. (P&D) Act, 1963 (exempted Development) Regs., 1967, No. 176 of 1967, Part III, class 6.
67. Information supplied by the Department of Agriculture.
68. Circular ENV. 5/72 (14/3/72) as amended by ENV. 23/73 (27/8/73).
69. See ENV. 23/73 P.3.
70. Dail Reports, Vol. 260, col. 542 (18/4/72).
71. Public Health Amendment, Act 1907, S 51.
72. P.H. (2) Act, 1878, S 129, As amended by P.H.A. Act 1907, S 51.
73. P.H. (2) Act, 1878, S 130.
74. See 3.1.2
75. See 1.6.2 and 3.1.8.
76. S 20.
77. See 1.5.2 to 1.5.4.
78. Circular L.8/1971. (10/3/71).
79. S.R.5: 1970.
80. HMSO, 1960.
81. *An Foras Taluntais,* by V. A. Dodd.
82. Circular L.8/1971. (10/3/71).
83. See Report on Water Pollution, Appendix 9.
84. Para. 9-4-1.
85. Para. 9-5-1.

86. See 1.4.2 and 1.6.2.
87. See 1.6.2.
88. See 1.6.2.
89. L.R. 2 Ch. App. 478.
90. See 1.6.2.
91. S 16.
92. S 16.
93. S 11.
94. S 27.
95. S 27.
96. S 6.
97. See also section 5.
98. Merchant Shipping (Safety Convention) Act, 1952, S 10, as extended by the Merchant Shipping Act, 1966, S 9. See the Merchant Shipping (construction) Rules, 1953, SI No. 339 of 1953 and Merchant Shipping (Passenger Ship Construction) Rules, 1967, SI, No. 98 of 1957.
99. Merchant Shipping Act, 1966, S 3. See the Merchant Shipping (Cargo Ship Construction and Survey) Rules, 1967, SI No. 99 of 1967.
and Merchant Shipping (Dangerous Goods) (Amendment) Rules, 1968, SI, No. 179 of 1968.
102. SI, No. 298 of 1936.
103. M.S.A., 1952, S 41.
104. M.S.A., 1952, S 42.
105. Merchant Shipping (Dangerous Goods) Rules, 1967, SI, No. 105 of 1967, and Merchant Shipping (Dangerous Goods) (Amendment) Rules, 1968, SI, No. 179 of 1968.
106. M.S.A., 1953, S 38.

# 4 Coastal waters

Ireland's coastal waters, running along 1738 miles of coastline, are, except for areas adjacent to major cities and industrial centres, relatively unpolluted.[1]

Controls over pollution of coastal waters are exercised almost exclusively by public authorities acting under statutory powers. These authorities are the Minister for Transport and Power, local authorities and harbour authorities. Apart from legislation controlling oil pollution, which is dealt with in the next section, the most important legislation enabling pollution control of coastal waters to be exercised is the Foreshore Act, 1933, which confers wide control powers on the Minister for Transport and Power. Unfortunately, this Act is lacking in that the definition of 'foreshore' for its purposes excludes the possibility of ministerial control over that part of the shore above high water mark of ordinary or medium tides. Local authority jurisdiction extends to the coast of its district. Harbour authorities have some powers to control pollution within the limits of their harbours.

The law controlling pollution of coastal waters is identical in many respects to that controlling pollution of inland waters. In this section it is proposed to indicate the applicability or non-applicability of controls over inland waters to coastal waters and to deal with provisions specifically applicable to coastal waters.

## 4.1 FIXED DISCHARGE POINTS

**4.1.1** Control over the siting of points of discharge of pollutants to coastal waters may be exercised by the same authorities as are mentioned in 3.2.2. It is relevant to note however that harbour authorities[2] have jurisdiction over the water areas of their harbours and land areas which are their own property. In the case of development adjacent to the foreshore whch could conceivably affect the interests of a harbour authority, it is usual practice for the planning

4.1.1 (continued) authority concerned to consult with the harbour authority prior to granting a permission.

In addition to the controls common to inland and coastal waters, additional control over the seashore and foreshore is exercisable by the Minister for Transport and Power by virtue of the Foreshore Act, 1933. For the purposes of this Act, 'foreshore' means 'the bed and shore, below the line of high water of ordinary or medium tides, of the sea and of every tidal river and tidal estuary and of every channel, creek, and bay of the sea or of any such river or estuary.'[3] Seashore means 'the foreshore and every beach, bank and cliff contiguous thereto, and includes all sands and rocks contiguous to the foreshore.'[4] Substantial stretches of the foreshore belong to the state, and the Minister for Transport and Power is empowered to purchase or lease by agreement any foreshore or right over same.[5] He may, if it is in the public interest, make a lease to any person of any state foreshore. Such a lease must contain a proviso for re-entry for breach of its terms and such covenants, conditions and agreements as the Minister thinks proper or desirable in the public interest and agrees with the leasee.[6] These leases may be used to control development and activities near coastal waters and have been so used. A relevant example of the use of such a lease to control pollution is the controversial lease of state foreshore made to the Gulf Oil Terminals (Ireland) Ltd. for the building of an oil terminal at Bantry Bay. The company was informed, prior to getting the lease of the implications for it of legislation enacted to prevent pollution, and was required to make satisfactory arrangements for safety of navigation, prevention of oil pollution and preservation of amenities. The lease eventually granted contained clauses designed to prevent pollution and an interesting proviso stipulating that, in the event of pollution control arrangements proving inadequate or defective, the Minister would have right to set up a harbour authority in the area should such action be necessary or desirable.[7] Similar

**4.1.1 (continued)** control may be exercised by virtue of licences which the said minister may grant (in the public interest) in respect of state foreshore authorising the licensee, *inter alia,* to 'place any material or to place or erect any articles, things, structures or works in or on such foreshore... or to use or occupy such foreshore for any purpose.' The licences must contain such covenants, conditions and agreements as the Minister thinks proper or desirable in the public interest and agrees with the licensee, and a power to the Minister to terminate the licence on breach, non-performance or non-observance of its terms.[8] Licences and leases may authorise the exploitation of minerals[9] in the foreshore to a depth of thirty feet.[10] The Minister consults all interested bodies when he receives an application for a foreshore lease or licence. He may refuse to approve maps, plans and specifications for the erection of structures on non-state owned tidal lands,[11] and may cause the removal of dilapidated structures on *any* foreshore on the grounds that such proposed structures or such structures respectively would cause an obstruction to navigation or fishing.[12] Structures unlawfully erected on state foreshore may be removed.[13] The Minister may, by regulation, control public use of state foreshore.[14]

## 4.1.2 Controls over processes

Controls over processes may be exercised by all the methods described or referred to in the preceding paragraph. The Planning Act, 1963, specifically empowers planning authorities to include in their development plans objectives to control pollution of the seashore.[15]

The Alkali, etc. Works Regulation Act, 1906, makes it an offence to discharge chemical waste from scheduled works, whether into coastal waters or otherwise, 'so as to cause a nuisance.'[16]

The Rivers Pollution Prevention Acts, 1876 and 1893, are not applicable in respect of pollution of coastal waters

4.1.2 (continued) as the Minister for Local Government has never made the order determining the geographical extent to which the sea and tidal waters come under the control of these Acts.[17] Furthermore, S 19 of the 1876 Act provides that no offence shall be committed by a sanitary authority which, in the exercise of a statutory power or duty, carries sewage into the sea or tidal waters. The Public Health (Ireland) Act, 1878, S 19 (requiring sanitary authorities to purify sewage before discharging it) does not apply in respect of discharges into coastal waters.[18] The nuisance provisions of the same Act do apply to ships and vessels in any river, harbour or district within the jurisdiction of a sanitary authority; and where any river, harbour or water is outside the jurisdiction of a sanitary authority, ships or vessels therein are, unless otherwise provided by the Minister for Local Government, deemed to be within the jurisdiction of the nearest sanitary authority.[19]

Section 47 of the Public Health Acts (Amendment) Act, 1890, does not apply in respect of discharges to coastal waters.

The Electricity Supply (Amendment) Act, 1945, S 42 does not, *prima facie,* extend to discharges to coastal waters unless, perhaps, a generating station is being constructed or operated on the banks of a river where such river enters coastal waters.[20]

The Fisheries Acts 1959 and 1962 apply to discharges into coastal waters as 'waters' for the purpose of these Acts include estuaries and any part of the sea.[21] In addition, special provisions are contained in the 1959 Act for the enforcement by sea fisheries protection officers of the provisions relating to pollution in the Act,[22] and for the exercise by district courts of jurisdiction in relation to offences under the Act.[23] Section 253 of the 1959 Act, although primarily intended to protect molluscs, is in effect, a prohibition of certain types of pollution because it

4.1.2 (continued) makes it an offence 'to deposit ballast, rubbish or other substance' or 'to place any implement apparatus or thing prejudicial, or likely to be prejudicial, to any oyster bed or oysters or brood or spawn thereof or oyster fishery, except for a lawful purpose of navigation or anchorage' within the limits of a licenced oyster bed. This section was extended to include mussel, periwinkle and cockle beds and fisheries by Section 282 of the said Act. An offender may be punished by fine and is liable to make good the full loss suffered by the licensee of such fisheries. Section 290 of the 1959 Act, as amended,[24] provides that it is an offence to discharge ballast from a vessel within any estuary, harbour or place unless such discharge is lawfully permitted. The penalty is a fine of up to £20.

Harbour authorities have jurisdiction over harbour areas, the limits of which are indicated in the Harbours Act, 1946.[25] These areas include a considerable amount of coastal waters. A harbour authority is obliged, *inter alia,* to take proper measures for the proper management control and operation of their harbour,[26] for the maintenance of all property and facilities under their control,[27] and for the cleaning, scouring, deepening, improving and dredging of their harbour and approaches thereto.[28] It has power to provide facilities at its harbour including storage facilities[29] and ballast;[30] to act as warehousekeepers;[31] to appropriate part of the harbour to the exclusive use of others; to remove obstructions within the limits of the harbour,[32] and to place and maintain buoys.[33] In addition, a harbour authority has a general power to make bye-laws 'for the good rule and government of their harbour' including bye-laws 'providing that the harbour master may remove nuisances from within the limits of the harbour.'[34] Unfortunately, harbour authorities have not made full use of their powers to control pollution other than oil pollution.[35]

Harbour masters must be notified of the arrival of vessels,[36] and may, subject to harbour bye-laws, give

4.1.2 (continued) directions to the masters of vessels for certain purposes, including the protection of persons or property and the regulation of traffic.[37]

Provisions in the Harbours Act, 1946 for ensuring the safety of a harbour or vessels therein include the requirement that hazardous goods including oil of vitriol be properly and distinctly marked;[38] that a harbour authority may prohibit the bringing of certain articles into the harbour,[39] and that ballast, earth, stones or other substances may not be discharged into harbour waters without authorisation from the harbour authority.[40]

These provisions of the Harbours Act may operate as direct or indirect controls on pollution of coastal waters.

Other controls over pollution of inland waters, not dealt with in this section, may, in certain circumstances, be applicable to coastal waters.[41]

## 4.1.3 Requirements as to pre-treatment of any discharge

See preceding paragraphs in this section and 3.1.3.

## 4.1.4 Qualitative and quantative limits of discharges

See preceding paragraphs in this section and 3.1.4.

## 4.1.5 Methods of discharge

See preceding paragraphs in this section and 3.1.5.

## 4.1.6 Monitoring to be done by discharger

See preceding paragraphs in this section and 3.1.6.

## 4.1.7 Enforcement process

The process by which controls over pollution of inland waters which also apply to coastal waters has been described in the preceding paragraphs in this section and in 3.1.7. The Foreshore Act, 1933, is enforced by the Minister for Transport and Power. Leases made under the Act may be enforced by:

(i) re-entry on the foreshore leased for breach, non-performance or non-observance of any covenant, condition or agreement;

(ii) action for breach of a term in a lease;

(iii) threat of non-renewal if the leasee appears to require an extension of the lease. (It is illegal for such leases to contain renewal clauses.)[42]

Licences granted under the Act may be enforced by:

(1) termination,

(2) action for breach of a term,

(3) threat of non-renewal. (It is illegal for such licences to contain renewal clauses.)[43]

The Act also makes it a general offence to throw into the sea adjacent to any tidal land 'any china, earthenware, metal or other article (whether whole or broken) which would or might cause injury to a person bathing... or using such land, nor any material substance (whether liquid or solid) which would or might be injurious or offensive to any such person.' The offence is punishable by a maximum fine of £5.

## 4.1.8 Individual powers

See 3.1.7. The Minister for Transport and Power may, if he thinks fit, hold a public inquiry before making a lease or granting a licence under the Foreshore Act, 1933. Public opinion or an interested individual might influence the Minister to hold such an inquiry.

## 4.2 CONTROLS OVER DUMPING IN OR NEAR COASTAL WATERS

The Foreshore Act, 1933, S13, prohibits the deposit, save in accordance with ministerial consent, or any material whatsoever on any foreshore or seashore or other place from which such material might, by the operation of natural causes, be blown or washed or moved to a foreshore. The maximum penalty is £10. A person convicted of such an offence may be ordered by the court to remove such material. Failure to comply with such order is an offence punishable by a fine of up to £10 plus £1 per day for a continuing offence. Section 14 of the Act prohibits the throwing, depositing or leaving on any tidal land of glass, china, earthenware, metal or other article (whole or broken) which might injure a person bathing in or using such land or any material or substance which would or might be injurious to any such person. The maximum penalty is £5. Other controls over dumping and run-off from land into coastal waters are discussed in Section 6.

1. See Report on Water Pollution, paras. 5.7.1 to 5.8.2.
2. See Harbours Acts, 1946, 1947.
3. S1.
4. Id.
5. S5.
6. S2.
7. Dail Reports, Vol. 262, Cols. 495-497 (4/7/72).
8. S3.

9. The word 'minerals' includes all minerals within the meaning of the Minerals Development Act, 1940, other than scheduled minerals, mineral compounds and mineral substances. M.D.A. 1940, S82.
10. S 2(7) and S 3(1).
11. S 10.
12. S 11.
13. S 12.
14. S 8.
15. See 3.1.2.
16. S 3.
17. See 3.1.2.
18. See 3.1.2.
19. S 126.
20. See 3.1.2.
21. Fisheries (Consolidation) Act, 1959, S 3 and see 3.1.2.
22. S 233.
23. SS 234-236.
24. Fisheries (Amendment) Act, 1962, S 32.
25. S 2(3) and S 47.
26. S 47(1).
27. S 47(2).
28. S 48.
29. S 50.
30. S 52.
31. S 54.
32. S 56.
33. S 58.
34. S 60 and second schedule.
35. See 5.1.1.
36. S 65.
37. S 67.
38. S 86.
39. S 87.
40. S 88.
41. See 3.1.2.
42. S 2.
43. S 3.

# 5 Seas

Ireland claims jurisdiction over the territorial sea to a distance of three nautical miles from the nearest point of baselines drawn along the coast in accordance with the provisions of the Maritime Jurisdiction Act, 1959,[1] and an order made thereunder.[2] In addition, the exclusive fishery limits of the state for the purposes of part XIII of the Fisheries (Consolidation) Act, 1959, extend to that part of the sea which lies at a distance of twelve nautical miles from the nearest point of the said baselines.[3] These sea areas are policed by four minesweepers and one fisheries protection vessel owned by the Irish 'navy'.

## 5.1 POLLUTION BY OIL

Controls over pollution of the sea by oil are exercisable to a greater or lesser extent by an enormous number of bodies. These include:

*The Minister for Local Government:* overall responsibility for clearance of oil pollution, control under the Planning Act, 1963;

*The Minister for Transport and Power:* control under the Harbours Acts, 1946, 1947; and under the Foreshore Act, 1933; and under the Continental Shelf Act, 1968;

*The Minister for Agriculture and Fisheries:* control under the Fisheries Act, 1959, 1964;

*The Ministry for Industry and Commerce:* control under the Oil Pollution of the Sea Acts, 1956, 1965; Minerals Development Act, 1940; the Petroleum and other Minerals Development Act, 1960; and the Continental Shelf Act, 1968;

*The Minister for Labour:* control under the Dangerous Substances Act, 1972;

5.1.1 (continued) *The Minister for Defence:* clearance of oil pollution at sea, reporting on oil pollution;

*Local authorities:* control under the Planning Act, 1963, the Dangerous Substances Act, 1972; responsibility for clearing oil pollution of the foreshore and immediately off the foreshore;

*Harbour authorities:* control under the Harbours Acts, 1946, 1947, the Oil Pollution of the Sea Acts, 1956, 1965, and the Dangerous Substances Act, 1972; responsibility for clearing oil pollution in their harbours.

Ireland has implemented the international conventions for the prevention of pollution of the sea by oil, 1954, and the 1962 amendment thereto,[4] and has signed a number of other conventions relevant to oil pollution control.[5] At the special assembly of IMCO in November, 1968, areas off Tuskar and Fastnet, off the Irish coast were amongst the areas for which traffic separation schemes in respect of sea traffic were approved.[6]

Arrangements for clearing up oil pollution of the sea are largely extra-legal.

## 5.1.1 Controls over discharges from ships and shore installations

The Oil Pollution of the Sea Act, 1956, and the Oil Pollution of the Sea (Amendment) Act, 1965, and regulations made thereunder provide a number of controls over the discharge of oil or oily mixtures (as defined in the Acts) from ships and shore installations.

The discharge of oil or oily mixtures from an Irish registered ship in prohibited zones (defined in the Act and extended by order)[7] is an offence for which the owner and master of the ship are liable.[8] A like discharge from a vessel or place on land or from an apparatus for transferring oil to

5.1.1 (continued) or from a vessel is also an offence for which the owner and master of the vessel, the occupier of the place on land and the person in charge of the apparatus, respectively, are liable.[9] The Acts provide certain defences to the aforesaid offences.[10] These offences are punishable by a fine not exceeding £500 on summary conviction, and £50 per day for a continuing offence; and by a fine not exceeding £5000 on conviction on indictment plus £500 per day for a continuing offence.[11] Masters of all Irish registered ships, and in special circumstances masters of any ship outside the prohibited zones, are required to keep prescribed records in which details of all discharges of oil and other specified matters must be entered.[12] The Minister has made regulations requiring the keeping of records of the transfer of oil to or from vessels in Irish territorial or inland waters.[13] Failure to keep such records or contravention of the regulations relating to such records is punishable by a fine of up to £250. False or misleading entries in, or alterations of, records is punishable by a fine of up to £250 and/or six months imprisonment.[14] Other regulations under the Act provide that Irish registered ships using oil as fuel for engines or boilers be fitted so as 'effectively to prevent oil fuel from leaking or draining from machinery spaces into bilges unless the contents of the bilges are subjected to an effective means of separating the oil therefrom before they are pumped into the sea.'[15] The aforesaid provisions are enforceable by the Minister for Transport and Power.

The Continental Shelf Act, 1968, S 7, extends the provisions of S 10 of the Oil Pollution of the Sea Act, 1956, as amended, to oil which enters the sea from a pipeline, or, 'otherwise than from a ship, as a result of any operation for the exploration of the sea bed and subsoil or the exploitation of their resources in the designated area.' It is a defence to prove that the escape of oil occurred without the permission of the owner of the pipeline or the person carrying on operations or that reasonable care was taken to prevent or stop it. The penalty is a fine not exceeding £100

5.1.1 (continued) on summary conviction, or such fine as the court thinks fit in the case of conviction on indictment. This Act is enforceable by the Ministry for Industry and Commerce.

**Harbours**

Harbour authorities have powers to control oil pollution within harbour limits.[16] The Petroleum Act, 1871, S 4, requires that every harbour authority where petroleum (as defined in the Petroleum Acts, 1871 to 1881) is landed submit bye-laws to the Minister for Transport and Power regulating the landing and other matters relating to petroleum. For the purposes of the 1871 Act, harbour includes, *inter alia,* any dock, pier, jetty or other works at which goods are shipped or unshipped. The Harbours Act, 1946, S 60, empowers harbour authorities to make bye-laws, 'for the good rule and government of their harbours.' Pursuant to these enabling sections, a number of harbour authorities have made bye-laws relating to petroleum in their harbours, as has, interestingly, Gulf Oil Terminals (Ireland) Ltd.[17] for petroleum operations on state foreshore leased to them under the Foreshore Act, 1933.[18] These bye-laws vary in detail and scope but they generally provide for the regulation of such matters as the discharging, handling, loading, storage and distribution of petroleum in harbours. Their effectiveness is, however, greatly diminished by the fact that the maximum penalty for breach of such bye-laws is only £10, and by the fact that many bye-laws are out of date and inapplicable to modern petroleum operations.

The Harbours Acts, 1946 and 1947, contain a number of provisions under which a harbour authority may exercise direct or indirect control over oil pollution in harbour waters.[19] These controls have been dealt with in sections 3.1.2 and 4.1.2.

The Oil Pollution of the Sea Acts, 1956 and 1965, make special provision for the exercise of control by harbour

5.1.1 (continued) authorities over oil pollution. A harbour master may specify where, at what times, and the conditions under which ballast water from a petroleum-carrying ship may be discharged in harbour water.[20] All discharges and escapes of oil or oily mixtures from vessels or land into harbour waters must be reported to the harbour master, as must the cause of such occurrences. Failure to make such a report is punishable by a fine of up to £100.[21] Oil may not be transferred to or from a vessel in a harbour at night unless notice of the prescribed transfer is given to the harbour master. Failure to give such notice is punishable by a £50 fine.[22]

The Dangerous Substances Act, 1972, requires that petroleum-spirit (as defined in the Act)[23] be kept in stores licensed by the proper local or harbour authority (or if the store is owned by such authority, licensed by the Minister for Labour). The keeping of a small quantity of petroleum is exempt from this requirement as is the keeping of petroleum-spirit by a carrier or other person for conveyance in conformity with the provisions of the Act or in the exercise of functions conferred on him by or under the Act.[24] Containers of petroleum-spirit must be marked in a prescribed way.[25] Accidents involving petroleum kept under licence which occur near or in connection with licensed premises and which involve injury to property must be notified to the Minister, as must accidents near or in connection with licensed premises and which involve injury to property must be notified to the Minister, as must accidents near or in connection with vessels or vehicles on which petroleum is being loaded, unloaded or conveyed.[26] The Minister may then direct a formal investigation of such accidents[27] or direct an inspector to make a report thereon.[28] Licences to store petroleum-spirit may be granted by the Minister for Labour or by a local or harbour authority. All three may grant or refuse such licences at their discretion, and may, if they think fit, publish particulars of the application for the licence and invite

**5.1.1 (continued)** representations concerning it. The Minister and local harbour authorities may attach such conditions as they think proper to any licence, but local and harbour authorities are subject to regulations when attaching such conditions.[29] If a local or harbour authority refuses to grant a licence or attaches conditions to a licence with which the applicant is dissatisfied, appeal lies to the Minister.[30] Appeal from the Minister's decision on appeal, or from his decision not to grant a licence or to attach conditions to it with which the applicant is dissatisfied lies to the High Court.[31] Contravention of any provision of the Act, or of any instrument thereunder, or any provision attached to a licence, is an offence.[32] The Minister or a local or harbour authority, as appropriate, may bring the prosecution.[33] The penalty is a fine of up to £100 on summary conviction, and/or six months imprisonment, plus £20 for each day in which the contravention is continued after conviction.[34] Regulations may also provide for the revocation of licences.[22]

Ships carrying petroleum-spirit must, on entering any harbour in the state, give notice of the nature of such cargo to the harbour master. Failure to give such notice is an offence for which the owner or master of the ship are liable. It is a defence for them to prove that they did not know, or could not with reasonable diligence have found out the nature of the goods in respect of which the proceedings were taken.[36]

**Other controls**

Further controls over discharges of oil into the sea are exercisable under various statutes already described in this report, and particularly under the Minerals Development Act, 1940, and the Petroleum and Other Minerals Development Act, 1960 (both of which were extended to the Continental Shelf area by the Continental Shelf Act, 1968, Section 4); under the Foreshore Act, 1933; and under the

**5.1.1** **(continued)** Local Government (Planning and Development) Act 1963.

**Equipment and arrangements in respect of oil pollution**

No statute specifically obliges any oil pollution control authority to provide equipment which would prevent or clear up oil pollution. The Harbours Act, 1946, provides that a harbour authority *may* provide for, or in connection with, their harbour such equipment as they think proper,[37] and that they may provide ballast.[38] The Oil Pollution of the Sea Act, 1956, S 13, provides that harbour authority may provide, or cause to be provided, facilities for the discharge of oily residue from vessels using the harbour and arrange for its disposal, and that the Minister for Industry and Commerce may direct the harbour authority to provide such facilities if he considers that such facilities are necessary at a particular harbour or that existing facilities are inadequate. The Oil Pollution of the Sea (Amendment) Act, 1965, S 5, provides that the said Minister, if of the opinion that facilities for the discharge of oil and oily residue from ships and arrangements for their disposal are necessary at any premises occupied by an oil refinery, shipbuilding or ship repairing yard, or that existing facilities at such premises are inadequate may require the owner of the refinery or yard to provide such facilities. Failure to comply with such requirements is an offence.

Statutory responsibilities in respect of oil pollution of the sea are confined to the administration of various Acts and have no relationship to the physical task of clearing up oil pollution. Such responsibility has been treated as an administrative matter. Following the report of the Committee on Oil Pollution;[39] detailed arrangements were made to deal with oil pollution.[40] At central level, the Minister for Local Government has overall responsibility for the clearance of oil pollution. Local responsibility rests with maritime local authorities and harbour authorities — local authorities being required to deal with oil pollution which has come ashore and oil slicks immediately offshore

**5.1.1** **(continued)** which pose a threat to their areas, and harbour authorities being responsible for clearing up oil pollution in their harbours. The Minister for Defence is responsible for clearing oil pollution in the open sea. There appears to be no statutory basis for the exercise of such function by local authorities. All local authorities concerned were:

(i) required to draw up contingency plans or joint contingency plans to deal with oil pollution. (Guidelines as to the contents of such plans were provided);

(ii) exhorted to cooperate with each other and with harbour authorities in the making of plans and in the administration of arrangements to deal with oil pollution;[41]

(iii) given detailed advice on methods of dealing with oil pollution;

(iv) advised to acquire equipment and dispersants for clearing up oil pollution;

(v) referred to relevant publications dealing with oil pollution; and

(vi) advised as to suitable materials and equipment which could be purchased, and of the address of suppliers thereof.[42]

A central liaison committee (representative of the Departments of Local Government, Defence, Agriculture and Fisheries, Transport and Power, and of the Institute of Engineers in Ireland and the County and City Managers Association), was established to provide guidance and advice to local bodies charged wth clearing oil pollution and to coordinate the efforts and arrangements of these bodies.

Arrangements have also been made for dealing with a major emergency, should it arise,[43] and for the detection and reporting of oil pollution.[44]

5.1.1 (continued) **Finance**

It has been agreed that the state will meet fifty percent of the cost to local authorities of clearance of oil pollution from beaches and immediately offshore, including purchase of approved dispersant, equipment and other materials necessary for such activity, subject to a limit on local authority expenditure for this purpose of a sum equal to the produce of 2.5p in the £1 on the rates in any year. All expenditure above this limit will be reimbursed by the State.[45] Financial arrangements with harbour authorities have not been finalised.[46]

**Personnel**

Oil pollution officers and deputy oil pollution officers have been appointed in areas covered by contingency plans. These officers are usually qualified engineers. A two-day conference for such officers was held in July, 1973.[47]

## 5.1.2 Facilities available at ports for the disposal of oily residues

Facilities at ports for the disposal of oily residues may be provided by harbour authorities, local authorities, private concerns and, at larger ports, by independent commercial operators. Administrative arrangements are being evolved whereby facilities and materials for dealing with oil pollution will be provided on a regional basis and coordinated by a central authority.[48] Facilities and materials available include those at the immediate disposal of a control authority and acquired specifically for dealing with oil pollution, and others which may be called upon if necessary. The following (provided for illustrative purposes only) shows the facilities and materials possessed solely for clearing up oil pollution by those harbour authorities who have replied to a circular seeking information on facilities available at their harbours:

5.1.2 (continued) *Cork Harbour Commissioners:* small stocks of oil dispersant and knapsack sprayers for dealing with minor oil spillages. (Major spillages are usually associated with nearby refinery traffic and the Irish Refining Company have two tugs and two launches equipped with oil dispersing equipment).

*Dublin Port and Docks Board:* stocks of emulsifier (BP 1100-X) and milled peat. The Board also possesses a seven hundred foot long boom which can be towed into position around an oil spillage by an inflatable dinghy equipped with an outboard motor.

*Galway Harbour Commissioners:* a minipack and one hundred gallons of dispersant.

*Limerick Harbour Commissioners:* one two-ton tipper trucks;
one hundred gallons of 'Polyclens' detergent;
one hundred gallons of 'Jaysol' detergent;
three hundred gallons of 'BP 1100-X' dispersant;
knapsack sprayers;
one hundred and fifty foot oil boom;
three quarters of a ton of milled peat.

*New Ross Harbour Commissioners:* none.

*Sligo Harbour Commissioners:* none.

Facilities and materials available in other ports are described in appendix III.

## 5.1.3 Powers to deal with or prevent accidental discharges

Statutes do not contain provision explicitly empowering any authority to deal with or prevent accidental discharges of oil into the sea. Administrative arrangements for clearing up oil pollution have been described in 5.1.1. All prevent discharges, including accidental discharges, in waters over which they exercise control. It must be noted, however, that the fact that a discharge was accidental may be a good defence to a charge of causing oil pollution.[49]

## 5.1.4 Enforcement

The enforcement of various provisions in Acts under which oil pollution control may be exercised has been described. The Oil Pollution of the Sea Acts, 1956, 1965 and the Dangerous Substances Act, 1972, provide for policing of the operation of these Acts by inspectorates, who have wide powers to carry out their functions.[50] Prosecutions can be brought by the appropriate ministers or control authorities. Information available on enforcement of oil pollution controls indicates that the Acts most frequently invoked are the Oil Pollution of the Sea Acts, 1956, 1965 and that the Cork and Dublin harbour authorities, as might be expected, are the most vigilant in enforcing these Acts. Prosecutions for oil pollution under the Acts were brought by harbour authorities as follows.[51]

| *Year* | *Control Authority* | *Number of Prosecutions* |
|---|---|---|
| 1970 | Cork Harbour Commissioners | 7 |
| | Dublin Port and Docks Board | 1 |
| 1971 | Cork Harbour Commissioners | 4 |
| | Dublin Port and Docks Board | 1 |
| 1972 | Cork Harbour Commissioners | 1 |
| 1973 | Cork Harbour Commissioners | 1 and 1 pending |

Prosecutions for offences under the Oil Pollution of the Sea Acts, 1956, 1965, were brought by the Minister for Transport and Power as follows:

5.1.4 (continued)

| *Year* | *Number of Prosecutions* | *Fines Imposed* |
|---|---|---|
| 1971 | 1 | £300 |
| 1972 | 2 | £900 |
| 1973 | 3 | *sub judice* |

Total fines of £1150, £1025 and £500 were imposed on offenders prosecuted by harbour authorities in the respective years. There are many instances of oil pollution in respect of which prosecutions are not taken, e.g. in the three year period, 1971-1973, there were twenty-four oil spillages in Cork Harbour, of which eleven caused residual pollution. The following table gives a breakdown of these spillages and of prosecutions taken by the Cork Harbour Commissioners, who appear to have the greatest problems with oil pollution: [52]

| *Year* | *Spills* | *Sources* | | | *Number of Prose-cutions* |
|---|---|---|---|---|---|
| | | *Vessels* | *Premises* | *Un-known* | |
| 1971 | 10 | 8 | 1 | 1 | 4 |
| 1972 | 8 | 6 | 1 | 1 | 1 |
| 1973 | 6 | 4 | 1 | 1 | 1 and 1 pending |

The Marine Rescue Coordination Centre at Shannon Airport acts as a communications centre for the receipt and dissemination of information concerning oil slicks at sea. There is no systematic reporting of oil pollution incidents which may occur at sea (apart from those affecting harbours), but airline pilots, vessels and the four minesweepers and one fisheries protection vessel in the Irish Navy all report instances of oil pollution of the sea.

### 5.1.5 Civil liability

Ireland has signed, but has not yet taken the necessary steps to implement, the international convention on civil liability for oil pollution damage, 1969. Enforcement of oil pollution controls has been by resort to criminal proceedings. This, of course, is unsatisfactory, as the penalties which may be imposed for oil pollution bear no relation to the cost of clearing up damage caused by such pollution.[53]

## 5.2 POLLUTION BY WASTES OTHER THAN OIL

### 5.2.1 Controls over dumping from ships and aircraft

Ireland has signed the international conventions on the prevention of dumping from ships and aircraft in the marine environment, 1972, and on the prevention of marine pollution by dumping of wastes and other matter at sea, 1972, but has not yet signed the international convention for the prevention of pollution by ships, 1973. Some controls may be exercised by harbour authorities under the Harbours Acts, 1946, 1947,[54] but it is difficult to see how any controls over such dumping could be enforced by other authorities who exercise control over the sea in view of the inadequate naval service at the disposal of the state.

### 5.2.2 Controls over pollution of the sea from sources on land

Ireland has not yet had an opportunity of signing the convention for the prevention of marine pollution from land-based sources, 1974. For controls exercisable, see 3.2. and section 6.

## 5.3 POLLUTION CONTROL RELATING TO THE CONTINENTAL SHELF

### 5.3.1 Geographical extent of jurisdiction

The rights of the state outside territorial waters over the seabed and subsoil for the purpose of exploring such sea bed and subsoil, and exploiting their natural resources are vested in the Minister for Industry and Commerce.[55] The Government has designated the areas within which the aforementioned rights may be exercised.[56] The distance from the coast to the outer limit of the total designated area varies from one hundred and ninety-five nautical miles in the south-west to twenty nautical miles in the east.[57]

### 5.3.2 Controls over seabed workings

Control over seabed workings in the continental shelf may be exercised by virtue of the Continental Shelf Act, 1968. This Act, *inter alia,* extended the application of the Minerals Development Act, 1940, and the Petroleum and Other Minerals Development Act, 1960,[58] to the continental shelf,[59] and it is by virtue of these Acts that the greatest control over seabed workings may be exercised. The Act also provided for the application of certain laws of the state to and for the exercise of jurisdiction by Irish courts over acts and omissions (which would be offences if committed in the state) in the continental shelf area.[60] It is an offence under the Act to construct, alter or improve any structure or works or to remove any object or material from a designated area without the consent of the Minister for Transport and Power.[61] The Minister for Industry and Commerce may, for the purpose of protecting installations in designated areas, make an order prohibiting ships from entering areas specified in the order without his consent. Entry without his consent is an offence.[62]

Offshore exploration activities in progress are solely concerned with finding oil and gas. Marathon Petroleum Ireland Ltd. holds exclusive facilities in 13,112 square miles of the offshore continental shelf[63] and is currently searching for oil and gas. This company holds eight petroleum leases and to date thirteen wells have been drilled in their leasehold areas. In another part of the designated continental shelf sixty-two holders of non-exclusive petroleum prospecting licences are engaged in general exploration activities. Permission has not been granted to any company to drill for oil and gas in territorial waters.[64]

## 5.3.3 Enforcement

Enforcement of control powers over activities in the continental shelf areas is primarily the responsibility of the Minister for Industry and Commerce. Information is not readily available on the extent of this enforcement, but it is understood that regular reports on the progress of exploration operations are furnished to the said minister. The minister is currently having all aspects of policy relating to petroleum exploration in both land and off-shore areas examined.[65]

1. Maritime Jurisdiction Act, 1959, SS 3, 4.
2. Maritime Jurisdiction Act, 1959 (Straight Baselines) Order, 1959, SI. No. 173 of 1959.
3. Maritime Jurisdiction Act, 1959, S 6(1) as amended by the Maritime Jurisdiction Act, 1964, S 2.
4. See Oil Pollution of the Sea Acts, 1956, 1965.
5. See 1-8.
6. Report on Oil Pollution, Prl. 1431, p.8.
7. O.P.S.A., 1956, schedule and Oil Pollution of the Sea (Extension of Prohibited Zones) Order, SI, No. 123 of 1957.
8. O.P.S.A., 1956, S 10, as amended by O.P.S.A.A., 1965, S 6.
9. O.P.S.A., 1956, S 11.
10. O.P.S.A., 1956, S 12, O.P.S.A.A., 1965, S 7.
11. O.P.S.A., 1956, S 23.
12. O.P.S.A., 1956. S 17 as extended by O.P.S.A., 1965, S 8.
13. Oil Pollution of the Sea Act (Records) Regs. 1957, SI. No. 206 of 1957, and Oil Pollution of the Sea (Records)(Amendment) Regs. 1967, SI, No. 124 of 1967.
14. O.P.S.A., 1956, S 17(4) and S17(5).
15. Oil Pollution of the Sea Act, 1956 (Ships' Equipment) Regs., 1957, SI, No. 280 of 1957, Article 3.
16. See 4-1-1.
17. See 4-1-1.
18. See Dail Reports, Vol. 260, Col. 25 (12/4/72).

# 6 Deposit of wastes on land

Responsibility for the control of deposits of wastes on land lies almost entirely with local authorities in their various capacities. With some notable exceptions in the larger cities, arrangements made for the disposal of refuse throughout the country are in many cases grossly inadequate. The law governing such matters is highly fragmented and is contained in a large number of statutes dating from the late nineteenth century.

## 6.1 CONTROL OVER THE DISTRIBUTION OF PRODUCTS AND MATERIALS

There are no controls over the distribution of products and materials particularly liable to cause disposal and pollution problems. Local authorities have power to exercise some control over such matters by the making of bye-laws.[1] There are no controls, as yet, over the disposal of toxic wastes.

## 6.2 CONTROLS OVER THE DEPOSIT OF WASTES—GENERAL CONTROLS

The Local Government (Planning and Development) Act, 1963, S22 places a duty on planning authorities to take such steps as may be necessary for securing the objectives which may be contained in the provisions of their development plans. Since purposes for which objectives may be indicated in development plans include provisions for prohibiting, regulating and controlling the deposit and disposal of waste materials and refuse[2], planning authorities are therefore expressly empowered to provide for such matters. Controls over the deposit of waste by any development subject to the control of the Planning Act may be, and are, exercised by attaching conditions to planning permissions and licences granted under the Act.

**6.2** **(continued)** The Public Health (Ireland) Act, 1878, S55, empowers sanitary authorities to provide fit buildings or places for the deposit of any matters collected by them in pursuance of part II of that Act. This section is imperative but there appears to be no obligation on sanitary authorities to provide places for the deposit of trade wastes with which part II of the said Act is not concerned.

Local authorities have powers to make bye-laws for the good rule and government of their areas and for the prevention and suppression of nuisances.[3] Bye-laws relating to litter have been treated as relating to 'good rule and government.'[4] Sanitary authorities may make bye-laws relating to many matters, for the most part nuisances, under the Public Health (Ireland), Act, 1878.[5]

**Domestic refuse**

The Public Health (Ireland) Act, 1878 provides that a sanitary authority may, and if required by the Minister for Local Government must, undertake or contract for the removal of house refuse.[6] Once such service has been undertaken or contracted for, a sanitary authority will be guilty of an offence if, having been required by notice from the occupier of any house to remove such refuse within seven days, it fails without reasonable excuse to perform or have that service performed.[7] Where a sanitary authority does not contract for or undertake the removal of house refuse, they may make bye-laws imposing the duty of such removal on the occupier of premises.[8] The deposit of domestic waste may also be controlled by the Planning Act.

**Trade refuse**

The Public Health Acts Amendment Act, 1907 provides for the removal of trade refuse by local authorities when required to do so by the owner or occupier of any premises, subject to a reasonable payment for such

6.2 (continued) service.[9] Local authorities may also make bye-laws controlling any trade which is an offensive trade under S128 of the Public Health (Ireland) Act, 1878,[10] in order to prevent any noxious or injurious effects of such trade.[11]

**Industrial refuse**

The deposit of wastes from industrial processes may be controlled as described in 3.1.2.

**Litter**

The Local Government (Planning and Development) Act, 1963 makes it an offence for any person to throw down, place or leave in or on any public place:

(a) any food remnants, orange peel, banana skin or other organic matter (whether waste or animal) or,

(b) any rubble, old metal, glass, china, earthenware, tin, carton, paper, rags or other rubbish, so as to create or tend to create litter.[12]

The penalty is a fine not exceeding £10. The Act provides that this offence may be prosecuted by the planning authority in whose area the offence is committed, and that any duly authorised officer of a planning authority may demand the name and address of any person whom he reasonably believes to be guilty of such offence. Where the person concerned does not give his name and address he is liable on conviction to a fine of £5. The creation of litter is also any offence under anti-litter bye-laws which have been made by a number of local authorities.

Anti-litter laws are not rigorously enforced. In the twelve months from June 1972 to June 1973 Dublin Corporation brought only two prosecutions against peole creating litter. Prosecutions were secured in both cases. In 1971 Cork Corporation initiated nine similar prosecutions

6.2 (continued) and convictions were secured in seven cases. Local authorities are generally reluctant to resort to prosecution for creating litter and prefer to rely on public cooperation and publicity campaigns for controlling litter. The futility of this reliance is universally obvious.

Sanitary authorities have power to provide receptacles for the temporary deposit of rubbish.[13]

**Street cleansing**

The Public Health (Ireland) Act, 1878 provides that local authorities may, and, if required by the Minister for Local Government, must undertake or contract for the cleansing of streets, and that they may also undertake or contract for the proper watering of streets in their district.[14] Where a sanitary authority does not undertake or contract for the cleansing of streets adjoining premises, they may make bye-laws imposing such duty on the occupier of such premises.[15]

**Abandoned vehicles**

The Road Traffic Act, 1961 empowers the Minister for Local Government to make regulations for the removal, storage and disposal of vehicles which have been or appear to have been, abandoned on a public road, or in a public car park, or which have been parked illegally.[16] Accordingly the Road Traffic (Removal, Storage and Disposal of Vehicles) Regulations[17] empower a road authority[18] to remove, store and dispose of a vehicle which has been abandoned on a public road or in a public car park or which appears to be so abandoned. Implementation of these Regulations is the responsibility of the Garda Siochana or the appropriate road authority. Furthermore, where an abandoned vehicle causes danger or obstruction or is illegally parked, the Garda may prosecute the person who abandoned it.

6.2 (continued) As regards the use of privately-owned land for the dumping of vehicles or vehicle parts, it is expressly laid down in the Local Government (Planning and Development) Act, 1963 that where land is used for such a purpose the use of the land is materially changed and planning permission for the new use is accordingly required. If such permission were not sought or obtained, the planning authority concerned could have recourse to their enforcement powers under the Act to secure the discontinuance of the use.

**Mining waste**

The Minerals Development Act, 1940, and the Petroleum and Other Minerals Development Act, 1960[20] contain a number of provisions by which extensive controls over waste from mines may be exercised. These controls include:

(i) powers to attach conditions to licences, leases, permits and permissions granted under the said Acts;

(ii) provisions requiring the payment of compensation for damage caused to the surface of land or to mineral deposits or water supplies, or for causing a nuisance, by mining or petroleum operations;

(iii) provisions requiring that operations be carried on so in such a manner as not to interfere with the amenities of the locality near the mines, etc.;

(iv) provisions empowering the Minister for Industry and Commerce to make regulations in respect of, *inter alia,* the disposal of waste products. Furthermore, mineral extraction and ancillary operations, such as the erection of buildings and tipping of waste, come under planning control.[21]

## 6.2.1 Controls over places where waste may be deposited

Local authorities have power to provide dumping grounds for refuse and to make bye-laws which may control, *inter alia,* the places where wastes may be deposited, and the type of waste which may be deposited.[22] The selection of suitable places for such dumping is a matter within the discretion of each local authority. Otherwise provisions in various statutes and other less formal methods of pollution control may operate to prevent the deposit of wastes or of specified types of wastes in certain places. Statutes controlling the deposits of wastes include the:

Rivers Pollution Prvention Act, 1876, SS 2, 4;
Public Health (Ireland) Act, 1878, S 77;
Public Health Acts Amendment Act, 1890, S 47;
Alkali, etc Works Regulation Act, 1906, S 3;
Foreshore Act. 1933, SS 13, 14;
Water Supplies Act, 1942, S 20;
Electricity (Supply) (Amendment) Act, 1945, S 42;
Harbours Act, 1946, S 88;
Fisheries Acts, 1959, SS 171, 172;
Continental Shelf Act, 1968, S 4.

The above statutes provide for controls over the deposit of wastes in or near waters.[23] Other statutes by which general controls over where and what types of waste may be deposited include the Minerals Development Act, 1940, the Petroleum and Other Minerals Development Act, 1960[24] and the Local Government (Planning and Development) Act, 1963.[25] Agricultural waste may be controlled as described in 3.1.2.

## 6.2.2 Controls over the method disposal of waste

Controls over the ways by which waste is to be disposed may be exercised by a number of authorities and methods dealt with throughout this report. Control over the (continued) disposal of waste collected by local authorities is a matter within the discretion of each local authority. There is widespread and justifiable dissatisfaction with the primitive methods employed by the vast majority of local authorities in disposing of refuse. A circular from the Minister for Local Government to all sanitary authorities has underlined the need for 'a basic review of existing practices' and has called for, and given advice as to the proper disposal of refuse.[26] There are no statutory controls over the methods by which local authorities dispose of rubbish apart from a requirement in the Public Health (Ireland) Act, 1878, S 55 that local authority dumps are not to create a nuisance. Common law rights may be exercised by an individual who is adversely affected by the dumping of refuse.[27]

## 6.2.3 Restoration land on which refuse has been deposited

See 6.2.2. The Minister for Local Government has advised sanitary authorities on the selecting of tip sites and on the treatment of land on which refuse has been deposited.[28]

See preceding paragraphs.

1. See 6-2.
2. Third Schedule, Cl.11.
3. Municipal Corporations (Ireland) Act, 1840, S 125. Local Government (Ireland) Act, 1898, S 16.
4. H.A. Street, *Local Government,* p.15.
5. See below.
6. S 52. House refuse means such as arises from inhabitancy or domestic occupation only. *Lyndon* v *Standbridge,* 26 L.G.ex. 386.
7. S 53.
8. S 54.
9. S 48.
10. See 3-1-2 Offensive Trades
11. S 51.
12. S 52.

13. Public Health (Ireland) Act, 1878, S 55.
14 Id., S 52.
15. Id., S 54.
16. S 97 as substituted by the Road Traffic Act, 1968, S 63.
17. SI No. 5 of 1971.
18. 'Road authority' means (a) the council of a county (b) the corporation of a county or other borough or (c) the council of an urban district. (R.T.A. 1961, S 3).
19. S 3.
20. See also 2-1-1.
21. L.G. (P&D) Act, 1963, S 3.
22. See 6-2.
23. See 3-1-2 and 4-2.
24. See 2-1-1 and 6-2.
25. See 1-4-2.
26. ENV. 2/74 (8/1/74). For a copy of the circular see the Appendix.
27. See 1-6-2.
28. ENV. 2/74 (8/1/74).

# 7 Noise and vibration

There are very few statutory controls over noise and vibration in Ireland. There are no independent national or local monitoring services. Such controls as do exist are vague and, for practical purposes, almost unenforceable. Except in the case of noise caused by aircraft, there are no specific standards set for noise limits. The Institute for Industrial Research and Standards is of the opinion that 'there are many industrial workers in this country who are exposed to noise environments which by widely accepted standards, would be injurious to their health.'[1] Greater control over noise and vibration is imperative.

## 7.1 STATIONARY SOURCES

### 7.1.1 Controls over land-use

Control over the siting of development which might create or be subject to excess noise or vibration may be exercised in two ways:

(i) by land-use zoning in development plans, and
(ii) by the refusal of permission to develop or by attaching conditions to same.[2]

There have been cases where permission to develop has been refused on the grounds that the development in question would be exposed to excess noise and vibration caused by traffic on nearby roads. The noise level standards in the latter case were deemed to be unacceptable because they exceeded seventy decibels for ten per cent of the time. In the planning permission granted to Tara Mines Ltd., conditions were imposed requiring that noise levels conform to internationally recognised standards.[3] The loudest noise level permissible by this development is seventy-five decibels during the day and sixty-five decibels at night, while vibration from blasting may not exceed 0.3 inches per second peak velocity.[22]

7.1.1 **Entertainment**
(continued) The Public Health Acts Amendment Act, 1890, S51, requires that houses, rooms, gardens or other places kept or used for public dancing, singing, music or other public entertainment of a like kind be licenced for such purposes. Licences may be granted upon such terms and conditions and subject to such restrictions as the District Court determines. Breach or disregard of the aforesaid terms or conditions is punishable by a penalty not exceeding £20, plus a daily penalty of £5 for a continuing offence and the licence may also be liable to be revoked. The Public Dance Hall Act, 1935, exercises somewhat similar control over public dancing with one major difference, i.e. that any interested individual or any member of the Garda Siochana has a right to object to the granting of public dancing licences.[5] Noise caused by public dancing is one of the most common grounds for such objections.

**Mining operations**
There is a dual system of control over noise caused by mining operations, viz, control under the Planning Act, 1963,[6] and controls under the Minerals Development Act, 1940, and the Petroleum and Other Minerals Development Act, 1960.[7]

**Limitations on land use in noise affected areas**
See preceding paragraph. The Institute for Industrial Research and Standards has carried out a survey on noise levels at Dublin Airport[8] and in the surrounding areas. This survey has been used for zoning purposes by Dublin Corporation. The Minister for Transport and Power is empowered to make orders imposing restrictions on the use of land in the vicinity of aerodromes including restrictions prohibiting building, or limiting the use of structures in specified areas near aerodomes.[9]

## 7.1.2 Controls over design and construction of noise generating plant and equipment

There are no specific statutory controls over the design and construction of noise generating plant and equipment. The Industrial Development Authority has an implied right to require that plant and equipment be so designed as not to create excess noise,[10] and the Minister for Labour, if satisfied that any manufacture, machinery plant, equipment, appliance, process or description of manual labour is of such a nature as to cause risk of bodily injury to persons employed, may, after consultation with the Minister for Health, make such special regulations as appear to him to be reasonably practicable and to meet the necessity of the case.[11] This power also exists in relation to work carried out on ships[12] and to building operations.[13] Despite a proven need for such regulations in respect of some factories,[14] the Minister has not yet exercised his power to control noise from plant and equipment in these factories.

Other statutory provisions relating to noise described in this section may indirectly operate as controls over the design and equipment of noise generating plant and equipment.

## 7.1.3 Restriction on emissions of noise and vibration

The Local Government (Planning and Development) Act, 1963, S 51, provides that it is an offence for any person

'(a) in any public place or in connection with any premises which adjoins any public place and to which the public are admitted[15] or

**7.1.3 (continued)** (b) upon any other premises either:

(i) by operating, or causing or suffering to be operated any wireless, loudspeaker, television, gramophone, amplifier, or similar instrument, or any machine or other applicance or
(ii) by any other means to make or cause to be made, any noise or vibration which is so loud, so continuous or so repeated, or of such duration or pitch or at such times as to give reasonable cause for annoyance to persons in any premises in the neighbourhood or to persons lawfully using such place.'

Proceedings may not be taken by any person for an offence in respect of any premises referred to in S 51(1)(b) unless the annoyance is continued after the expiration of seven days from the date of a notice alleging annoyance, signed by not less than three persons residing or carrying on business within the area in which the noise or vibration is felt.[16] Contravention of the section is punishable by a maximum fine of £10. There are saving clauses for noise caused by aircraft or by statutory undertakers[17] and in proceedings brought in respect of noise or vibration caused in the course of a trade or business or in performing any statutory functions, it is a good defence for the defendant to prove that the best practicable means have been used for preventing or counteracting the effect of the noise or vibration.[18]

While there are no statutory provisions with respect to noise levels in workplaces, the Department of Labour issues guidlines on maximum noise levels for employed persons exposed to noise. These are purely advisory in nature and are not legally enforceable.[19]

### 7.1.4 Enforcement process

The process by which controls mentioned in this section are enforceable has been described.[20] The Factories Act, 1955, is enforceable by the Minister for Labour and by the Industrial Inspectorate. Information is not readily avaiable on the extent to which noise controls are enforced.

### 7.1.5 Individual rights

In addition to statutory rights already described or referred to, an individual may be able to sue for negligence, nuisance, trespass or under the rule in *Rylands* v. *Fletcher* if he has been exposed to excess noise.[21]

**Proposed new legislation**
The Local Government (Planning and Development) Bill, 1973, if enacted, will give planning authorities express power to reduce or prevent the emission or intrusion of noise or vibration.

## 7.2 MOBILE SOURCES

### 7.2.1 Motor vehicles

**Control over design and construction**[22]
The Road Traffic (Construction Equipment and Use of Vehicles) Regulations 1963,[23] as amended, require that every vehicle be fitted with an audible warning device but, subject to certain exceptions, gongs, sirens and other strident-toned devices are forbidden.[24] They also require that every vehicle be fitted with an exhaust silencer or other suitable device for reducing noise to a 'reasonable level.'[25]

7.2.1 (continued) **Controls over maintenance**

The aforesaid Regulations require that every vehicle and parts and equipment thereof (which includes silencers and similar contrivances) be maintained in good and efficient working order. The alteration of a silencer in such a way as to increase noise is an offence.[26] All parts of a vehicle subject to severe vibration and all parts relevant to the control of a vehicle are required to be efficiently fastened so as to prevent their working or coming loose.[27]

**Controls over use**

Vehicles — whether mechanically propelled or not[28] — must not be used on public roads while there are attached to the vehicles a public address system incorporating a loudspeaker or similar device. Exceptions are made and there is provision for local authority licensing of such use of vehicles.[29] Vehicles may not be used in a public place so as to cause excessive noise which could have been avoided by the use of reasonable care[30] nor may audible warning devices be sounded between 23.00 hrs. and 7.00 hrs. in certain areas.[31] The use of any vehicle without an exhaust silencer or other similar contrivance is an offence.[32] Races on public roads are not permitted without the prior consent of the appropriate local authority.[33]

## 7.2.2 Aircraft

The Local Government (Planning and Development) Act, 1963, S51 specifically exempts noise and vibration caused by aircraft.

Ireland has accepted noise standards for aircraft agreed upon at the International Civil Aviation Organisation meeting in Montreal in 1970. The Air Navigation (Noise Certificate and Limitation) Order, 1972,[34] and the Air Navigation (Noise Certificate and

7.2.2 (continued) Limitation) (Amendment) Order, 1973 [35] contain a number of controls over noise caused by subsonic aircraft. Noise abatement procedures are followed and scheduled flights of jet aircraft are not permitted between 23.00 hrs. and 7.00 hrs. but charter flights are allowed.[36] Runway selection takes into account the need to avoid populated areas. These controls are enforceable by the Department of Transport and Power.

There is no special legislation in force in Ireland at present concerning supersonic flights of civil and military aircraft. Such legislation has been drafted and can be introduced if necessary.[37]

Restrictions on actions for trespass and nuisance caused by aircraft noise are contained in the Air Navigation and Transport Act, 1936, as amended.

No action shall lie in respect of trespass or in respect of nuisance, by reason only of the flight of aircraft over any property at a height above the ground, which, having regard to wind, weather and all the circumstances of the case is reasonable, or the ordinary incidents of the flight, so long as the provisions of part II of this Act and any order made under the said part II and any regulations made by virtue of any such order are duly complied with.

## Notes

1 *Technology Ireland,* September, 1970.
2 See 1.4.2 and under *proposed new legislation.*
3 It is understood that the standards generally referred to in this country are those contained in I.S.O.: 0i996 *Assessment of Noise with regard to Community Response.*
4 Reported in the *Irish Times,* 26/3/74.
5 *Dunne Ltd.* v *Dublin County Council* [1974]1.1.R.45, at p.53.
6 PDHA, 1935, S 2(3).
7 See above.
8 See 2.1.1.
9 Air Navigation and Transport Act, 1950, S 14.
10 See 1.5.3.
11 Factories Act, 1955, S 71.
12 Id., S 87.
13 Id., S 88.
14 See Report of the Industrial Inspectorate, 1971.
15 This sub-section does not apply to public meetings.
16 L.G. (P&D) Act, 1963, S 51(3).
17 Id., S 51(5) (a) & (b).
18 Id., S 51(6).
19 Dail Reports, Vol.254, Col.1877 (17/6/71).
20 See 1.4.2, 1.5.3, 1.6.2, 2.1.1, 2.1.7.
21 See 1.6.2.
22 See 2.2.1.
23 SI No.190 of 1963.
24 Id., reg. 28.
25 Id., reg. 29.
26 Id., reg. 34.
27 Id., reg. 34.
28 Road Traffic (Construction Equipment and Use) Amendment Regulations, 1965, SI No.79 of 1965.
29 Road Traffic (Construction, Equipment and Use of Vehicles) Regulations, 1963, Reg. 35.
30 Id., reg. 85(2).
31 Id., reg. 85(3).
32 Id., reg. 85(2).
33 Id., reg. 88.
34 SI No.280 of 1972. See appendix II for a copy of this order.
35 SI No.187 of 1973. See appendix II for a copy of this order.
36 See Dail Reports, Vol.241, Col.2293 (30/10/69).
37 Dail Reports, Vol.260, Col.542 (19/4/72).

# 8 Nuclear energy

## 8.1 RADIOACTIVE SUBSTANCES

Three ministers are empowered to exercise controls over radioactive substances, viz, the Minister for Health has power to make regulations in the matter of medical radioactive substances under the Health Act, 1956, the Minister for Transport and Power has control powers under the Nuclear Energy Act, 1971,[1] and the Minister for Labour has control powers under the Factories Act, 1955, and the Dangerous Substances Act, 1972.

The Minister for Health has not made any regulations under the Health Act, 1953. He has instead proceeded by way of voluntary controls operated on his behalf by the National Radiation Monitoring Service. The main activity of this body has been the provision of a film badge service for continuously measuring the radiation dose received by those working with irradiation apparatus. It also advises hospitals, dentists and commercial enterprises on the use of such apparatus and on safeguards necessary for the protection of workers and of the public from radiation spillage.

The Minister for Labour has made two sets of regulations — The Factories Ionising Radiations (Sealed Sources) Regulations, 1972,[2] and the Factories Ionising Regulations (Unsealed Radioactive Substances) Regulations, 1972,[3] in the exercise of powers conferred on him by Sections 6, 20 and 71 of the Factories Act, 1955. These regulations prescribe measures to be taken to ensure the adequate protection against radiation of persons employed in factories and other places to which the Factories Act, 1955 applies. It is estimated that there are about 40 factories in Ireland to which the sealed sources regulations apply and that there is very little use of unsealed sources in Irish industry at present.[4]

8.1 (continued) The Meteorological Service of the Department of Transport and Power carries out regular measurements of the radioactivity of precipitation, settled dust and airborne particles in a number of selected locations throughout the country. In addition, measurements of strontium-90 activity are made on milk samples supplied by the Department of Health. The National Radiation Monitoring Service undertakes the monitoring of radioactivity in the Irish Sea through the monitoring of samples of seawater, fish and seaweed from selected sites off the Irish Coast.

## 8.2. NUCLEAR INSTALLATIONS

In November, 1973, the Government decided in principle that the Electricity Supply Board [5] should proceed with proposals for the construction of the first nuclear power station in Ireland.[6] The ESB proposes to locate the station at Carnsore Point in County Wexford.

It is believed that the ESB is considering the use of an American model called the light-water reactor, which would use ordinary water as a moderator element and that the station will be in the 650 MW range. It is anticipated that the station will be in operation by the 1980s.

The only legislation directly relevant to nuclear energy is the Nuclear Energy Act, 1971, and the Nuclear Energy Act, 1973, (Commencement) Order, 1973,[7] made thereunder establishing a Nuclear Energy Board, the functions of which are, at present, purely advisory. The Act is enforceable by the Minister for Transport and Power who acts on the advice of the Nuclear Energy Board.

## 8.2.1 Control over siting of the installations

This primarily a matter for the ESB but, as a state-sponsored body, the ESB is working very closely with the Minister for Transport and Power, who is guided in nuclear matters by the Nuclear Energy Board, the functions of which include advising the said Minister and any other Minister of State on nuclear energy and matters connected therewith[8] including all aspects of the location of nuclear installations.[9]

Other laws regulating land use will also control the siting of the proposed nuclear installations.[10]

## 8.2.2 Control over design and construction

One of the express functions of the Nuclear Energy Board is 'to advise the Government or, with the consent of the Minister for Transport and Power, any other person or group of persons engaged in training or research in nuclear science on the acquisition of nuclear reactors or radioactive devices for training and research purposes and, in the event of acquisition of such reactors or devices, on all aspects of their location, installation, operation and supervision,'[11] 'to advise the Minister on proposals for the construction of nuclear power stations and on all aspects of the installation, operation and supervision of such stations.

## 8.2.3 Controls over operation and maintenance

See preceding paragraph. Controls over the maintenance and operation of nuclear stations do not exist at present. The Nuclear Energy Board has the functions of preparing draft safety codes and regulations dealing with fissile fuel or other radioactive substances or devices and irradiating apparatus with due regard to standards recommended by international bodies dealing with nuclear energy.[22] The

Minister is empowered to assign further functions to the Board relevant to the operation and maintenance of nuclear installations.[13]

### 8.2.4 Controls over waste disposal

It is envisaged that control over waste disposal will be exercised in two ways, viz.

(i) by ministerial order,

(ii) by licences to which conditions may be attached.[14]

Other control authorities may also impose conditions as to waste disposal.[22]

### 8.2.5 Monitoring

There is no express provision for monitoring in the Nuclear Energy Act, 1971. Presumably this is one of the matters which may be required by provisions relating to the operation of nuclear installations.

### 8.2.6 Enforcement

The Nuclear Energy Act, 1971, will be enforced by the Minster for Transport and Power and the Nuclear Energy Board.

### 8.2.7 Individual Rights

The Nuclear Energy Act, 1971, does not contain any provisions on individual rights.

1. See 8.2.
2. SI No. 17 of 1972.
1. SI No. 249 of 1972.
4. *Technology Ireland,* November, 1973, pp.20-21.
   See also Reports of the Industrial Inspectorate, 1971, 1972.
5. See 1. 6.1.
6. Dail Reports, Vol. 272, Col. 1003 (8/5)74).
7. SI No. 319 of 1973.
8. N.E.A., 1971, S 4(1)(a).
9. Id., S 5(1)(a).
10. See 1. 4.2 and 3. 1.1.
11. N.E.A., 1971, S 5(1)(a).
12. Id., S 5(1)(c).
13. Id., S 5(2).
14. Id., S 6.
15. See 2. 1.2, 3. 1.2, 6. 2.

# 9 Products

Certain products, especially those used in agriculture, may, under certain circumstances, pollute the environment or have deleterious effects on humans. Control over these products is largely the responsibility of the Ministers for Agriculture and Health. Restriction on the use of products used in agriculture is, in practice, almost solely by restriction on distribution. This is so because from a marketing point of view, manufacturers regard it as essential to get the approval of the Department of Agriculture for products which they wish to sell to farmers.[1] The Department consults the research staff of An Foras Taluntais[2] before a decision is given. This system appears to be working well and in recent years manufacturers have withdrawn supplies and ceased generally to manufacture such products as aldrin, dieldrin and heptachlor, and they have restricted the supply of DDT.

Since 1964, An Foras Taluntais has been monitoring Irish foodstuffs for pesticide residues as part of its environmental research programme. This research, so far, has shown that the majority of samples tested contained very low residue levels, and, with one exception,[3] conformed both to British standards and to standards prescribed by the FAO/WHO Codex Alimentarius Commission.[4]

Advice on the actual use of products in agriculture is given to farmers by the agricultural advisory services[5] and in many radio and television programmes.

**Fertilisers, feeding stuffs and mineral mixtures**

The Fertilisers, Feeding Stuffs and Mineral Mixtures Act, 1955, regulates and controls the sale, manufacture for sale and other matters relating to these products.

Section 2 of this Act and the Fertilisers, Feeding Stuffs and Mineral Mixtures Regulations, 1957,[6] provide that the seller of the aforementioned products must give the purchaser thereof a written statement containing, *inter alia,* information on the chemical and other ingredients of fertilisers and on the nutritive and other ingredients of feeding stuffs and mineral mixtures.

Section 4 gives the Minister for Agriculture power to prohibit the sale or manufacture for sale of the products mentioned or of any specified class thereof.

Section 5 and the said regulations[7] require that all manufacturers for sale of fertilisers, mineral mixtures and many compound feeding stuffs be licensed by the Minister for Agriculture. The Minister may grant or refuse a licence at his discretion, may attach such conditions as he considers proper to any licence and may, at any time, by notice in writing:

(i) attach any condition to a licence, or

(ii) revoke or vary any existing condition, or

(iii) revoke a licence.

The number of licensed manufacturers on 31 March 1973 was:

| | |
|---|---|
| Fertilisers | 12 |
| Ground Limestone | 45 |
| Compound Feeding Stuffs | 230 |
| Mineral Mixtures | 23 |

From 31 March 1972 to 31 March 1973 legal proceedings were instituted in three cases where samples of feeding stuffs taken for analysis by authorised officers did not comply with statutory requirements in relation to quality.[8]

Section 6 empowers the said Minister to make certain regulations, including regulations:

(a) prohibiting, limiting or regulation the use of any specified article in the manufacture for sale of the aforesaid products,

(b) prohibiting the sale of any of these products under a specified description unless it conforms to a specified definition and standard,

(c) requiring manufacturers or sellers to disclose in a specified manner the presence of specified materials in these products. The Minister has, in fact, exercised his power under this section to prohibit certain antibiotics in compound feeding stuffs.[9] This Act is enforceable by the Minister for Agriculture and authorised officers.

Further control over certain agrochemicals (though not over many modern ones) is potentially exercisable by virtue of the *Poisons Act, 1961.* This Act established *Comhairle na Nimheanna* (The Poisons Council) which is obliged to include in its membership a person with special knowledge and experience of the use of poisonous substances in agriculture, and two farmers.[10] Under section 15 of this Act, the Minister for Agriculture is empowered, after consultation with, or on the recommendation of the council, to make regulations as to the use of poisons for agricultural and veterinary purposes. The Minister has not made such regulations. An amendment to the Act is contemplated which will enable the Minister for Agriculture to make regulations to control the use of pesticides if necessary.[11]

**Sheep dips**

Section 35 of the Diseases of Animals Act, 1966, and regulations made thereunder provide that only dips approved by the Minister for Agriculture may be used for dipping sheep for the prevention or treatment of sheep scab. When investigations carried out by An Foras Taluntais revealed that the kidney fat of slaughtered sheep contained dieldrin and aldrin residues, the Minister withdrew his approval of dieldrin and aldrin based dips in 1968. Dips containing these substances are no longer used in this country.[12]

**Weedkillers**

Certain weedkillers may be controlled by regulations made under the Poisons Act, 1961. The Poisons Act, 1961, (Paraquat) Regulations, 1968,[13] made by the Minister for Health, restrict the sale of paraquat to manufacturers and wholesalers of the product, chemists, druggists and persons licensed by local authorities. They also provide that paraquat be secured in sealed containers marked 'poison' and that cautionary instructions to be legibly marked on such containers. Health officers ensure that the regulations are observed.[14]

**Detergents**

No standards for the biodegradability of detergents have been prescribed for Ireland. Ireland has not signed the European Agreement on the Restriction on the Use of Certain Detergents in Washing and Cleaning Products, 1968.

**DDT**

There are no controls over the use of DDT in Ireland, nor are such controls likely in the near future.[15]

**EEC**

The Department of Agriculture is participating in discussions on draft proposals for Council Directives on pesticides.

1. Paper, *Agriculture,* by V. A. Dodd, at the Pharmaceutical Chemical and Allied Industries Association's Symposium on Environmental Pollution (10/1/74).
2. See 1-5-4.
3. See Sheep Dips.
4. Paper, *Man in Agriculture,* by D. W. Robinson, at the National Conservation Conference, 25/11/70.
5. See 1-4-5.
6. SI No. 264 of 1967.
7. Id., Reg. 13.
8. Annual Report for the Minister for Agriculture and Fisheries, 1972-73.
9. Animal and Poultry Compound Feeding Stuffs (Control of Antibioitics) Regs., 1972, SI, No. 335 of 1972.
10. S 4(1).
11. Dail Reports, Vol. 272, Col. 1494 (15/5/74).
12. See *Farm and Food Research,* May/June, 1969.
13. SI, No. 95 of 1968.
14. For further information see Dail Reports, Vol. 248, Col. 397 (2/7/70), Vol. 249, Col. 29 (28/10/70), and Vol. 261, Col. 1310 (13/6/72).
15. Dail Reports, Vol. 272, Col. 1494 (25/5/74).

# APPENDICES

## MEMORANDUM ON WASTE COLLECTION AND DISPOSAL

### Introduction

With some notable exceptions in the larger urban areas the hygiene and amenity standards of refuse collection and disposal throughout the country leave much to be desired. In the majority of cases this is the result of highly fragmented systems in which local areas are provided with a separate service of collection and disposal, often using small private contractors. Because of the economics of scale much local service can be provided cheaply only by collecting in local trucks or trailers, by tipping in an uncontrolled manner, and by offering a collection service of inadequate extent and frequency. The dumps and surrounding areas are evil smelling, subject to fires, breed flies, rats and other vermin and attract scavenging birds in large numbers. Not infrequently these dumps are within sight of the main road entering the towns. Such tips can be the cause of genuine public complaint. An unfortunate side-effect is that it is difficult to obtain suitable sites for new dumps because of the objections of local residents.

Acceptable standards can be achieved only by re-organising the service to a system based on the use of modern collection vehicles and on controlled tipping at suitable sites.

### Refuse collections

The modern collection vehicle has a totally enclosed refuse container. It incorporates a 'crush-pack' device which transfers the refuse from the receiving hopper to the container and at the same time compresses it, thus substantially increasing the capacity of the vehicle. Machines of this type are efficient and hygienic.

Manufacturers usually describe compression vehicles in terms of the volume of loose refuse which they estimate can be got into the vehicle by compression. Rated capacities of the larger vehicles in common use are in the 50-60 cubic yard range. The density of loose refuse has almost halved in the past thirty years due to the reduction of ash and cinder content and the increase in the amount of paper and plastic packaging and may at present be taken as averaging about 2.5 cwt per cubic yard.

From the limited data available the average weight of domestic and trade refuse collected by the public collection service appears to be in the range of 10-14 lbs per person per week. In counties operating a countrywide service with the large type vehicles described above an average of say 2.5 'fills' per day is found to be a normal output. Thus one of these machines would deal with the refuse from a population in the range of say 14,000-19,000 persons per five day week on the basis of a once weekly collection. For a twice weekly service the population served per vehicle would be substantially reduced owing to the additional collection time involved.

The vehicles need regular servicing and repair. It follows that in order to provide a dependable collection service an operational fleet of three or four vehicles as a minimum is required.

Crush-pack vehicles are available in a wide range of capacities and smaller versions may be suitable for collection purposes even in the smaller urban areas. In general, however, it appears that a strong case can be made for integrating urban areas in a country-wide service using the large capacity vehicles as basic units.

**Refuse disposal**

The objectionable dump conditions described above do not arise where the tip head is managed in accordance with the accepted principles of 'controlled tipping'. Controlled tipping or 'sanitary landfill' are terms used to describe a procedure for the dumping of refuse in an inoffensive manner, and which results in the reclamation of land for amenity or agricultural use.

The essential elements of controlled tipping are the deposition and compaction of refuse in shallow layers and the covering of the exposed surfaces with soil or other suitable inert material to form a 'seal.' 'Compaction' and 'cover' are intended to control odour emission, to prevent the disposal of light refuse by the wind, to prevent the breeding of flies on exposed putrescible matter, to facilitate rodent control and to reduce the risk of fires.

In order to operate controlled tipping it is necessary to have suitable machinery at the site to level and compact the refuse as it is received and to spread the covering material. It is also necessary to have an adequate supply of inert spoil on hand. On some sites a supply of suitable covering material can be obtained by excavating at the site. Otherwise suitable material must be imported. This might comprise road excavation material, demolition waste or material specially excavated elsewhere.

A suitable machine for control at a typical central dump would be a traxcavator of 1½ to 2 cubic yd. bucket capacity with alternative blade attachment. This machine can be used as a light bulldozer for handling and consolidating the refuse and covering material.

In order to make the operation of a properly controlled tip head economic, and to justify the manpower and machinery needed, a large quantity of waste must be brought to it. As a general rule it should be the aim to cater for a minimum population of the order of 20,000-30,000 at a central tip head. Such a population might produce about two hundred tons of waste per week. This amount would provide relatively light work for a machine of the kind referred to above, and a well used machine no longer suitable for heavy duty would suffice.

It follows from the foregoing that in the county areas the number of tip heads should be reduced substantially. In the average county only one or two central tip heads should be maintained for maximum efficiency. The use of large capacity crush-pack as recommended under *Refuse Collection* above will minimise haulage time and will facilitate the operation of central dumps on the most economical basis. Recommendations on the operation of controlled tipping are given in the appendix to this memorandum.

**Refuse disposal surveys and plans**

It is recognised that some local dumps will need to be maintained, for example, to facilitate private dumping. It would be best if such sites were developed as transfer stations from which the waste would be taken to a central controlled tipping site. In any case the standard of upkeep should be improved by providing adequate screens, by preventing fires, by introducing machinery and covering materials at frequent intervals and by taking suitable precautions to eliminate fly and rodent infestation. Local geographical and other considerations must influence the planning of service areas and site selection for the large central tip heads. In some cases it may be convenient for adjoining counties to consider joint arrangements.

County-wide surveys of refuse disposal needs should be initiated and schemes prepared for collection and controlled tipping based on the above criteria.

In carrying out their surveys the local authorities should include the quantities of waste produced from industrial and trade sources and should give consideration to accepting as many as possible of these wastes for disposal at their central tip heads on a fee charging basis.

Considerations might also be given to the possibility of developing a site suitable for disposal of hazardous or toxic wastes. This type of service would result in reduced numbers of uncontrolled private dumps and would tend to bring the disposal of difficult wastes under local authority control.

It is suggested that manpower requirements based on keeping the tip head open for deposit of waste from private sources outside normal working hours and at weekends. This would greatly reduce the offensive practice of fly tipping on roadsides and in waterways which is prevalent throughout the country.

**Selection of tip sites**

Various types of sites are suitable for controlled tipping including:

1. Mineral excavation: stone quarries, gravel pits, etc.

2. Low-lying land, marsh and bogland etc.

3. Valleys and ravines, disused railway cuttings etc.

In site selection a guiding principle should be the improvement of otherwise useless land. Thus sites which might be selected where low-lying land could be raised above flood level, dangerous or unsightly quarries or pits could be filled in or near towns, ravines and valleys could be filled to provide playing fields, parks etc. Much valuable reclamation work on these lines has been accomplished in some areas.

With the exception of the large centres of population there should be no great difficulty in obtaining suitable central sites, especially if it is known that the site is to be managed in accordance with the principle of controlled tipping. However, if no suitable site of the kinds mentioned above is available, it is preferable to select a site, suitably located and if necessary screened, comprising land which may have other possible uses, and may on that account be somewhat more expensive to acquire, than to settle for a more readily available site which may prove very difficult or expensive to manage to the necessary standard.

It is desirable that before tipping is commenced on any site all necessary site preparation should be completed and a working scheme drawn up. The working scheme should cover such matters as the provision of covering material, drainage details including any precautions necessary to avoid water pollution, the finished levels of the site, the direction of tipping operations and general site management. The planning section of local authorities might be consulted in the selection of sites and also asked to cooperate in preparing plans for amenity areas, etc. on the reclaimed land.

**Water pollution**

While it is not intended that this memorandum be regarded as a code of practice, the following comments on the dangers of water pollution by tip percolate are considered necessary. It is important in the selection and development of tip sites that provision be made for protection of ground and surface waters. Tipping should not be executed in such a manner that percolate can pass directly or through fissures to underground water sources or to streams.

In the first year after tipping the percolate from domestic refuse can have an organic load in the order of 6,000-7,000 mg/l, with a high concentration of sulphates. However, with the exception of sulphate content, the polluting load decreases with time and after three years it may be expected to be very low. In addition, if the percolate is derived from rain falling directly onto the tip area, and if tipping is carried out in a properly controlled manner, the actual quantity of leachate may be less than one third the total precipitation. Since, for instance, ten inches per annum of rainfall approximately equals six hundred gallons per acre per day the total percolate flow derived from rainfall may be expected to be quite small.

If the choice of tip site is such that ground or surface waters can flow laterally through the refuse the risk of water pollution is greater. The problem is even more acute when the refuse is tipped into standing water, and in this case precautions also may be necessary to prevent air pollution by odiferous gases, primarily hydrogen sulphide. If it is necessary to wet tip it is advisable to subdivide the flooded areas into separate lagoons and to pump dry each lagoon before tipping. If this is not practicable, it may be possible to wet tip with inert material, or to tip the refuse quickly in shallow layers to above the top water level, covering immediately with compacted inert soil. Wet tipping without offensive odours is generally possible for about seven months of the year, and this time should be availed of to prepare the site for dry tipping during the five warmer months. Where the danger of pollution of ground or surface waters does exist the following strategies may be contemplated:

Waterproof the base of the proposed tip area with puddled clay. Rapid tip in shallow compacted layers and provide final layer of water-proof earth cover, with good cross-fall to ensure steady run-off of rain. If the tip base cannot be waterproofed, provide a 2-3 ft thick layer of aerated medium such as gravel or clinker. If percolation from the tip cannot be avoided collect all leachate and discharge it to a foul sewer or treatment plant, or spray it over the surface of the tip.

Where toxic materials are to be disposed of much stricter precautions are necessary so that there is no possibility of entry of these materials to ground or surface waters.

**Bibliography:**

1. *Refuse Disposal* — Report of the Working Party on Refuse Disposal, Department of the Environment, HMSO London, 1971.

2. *Disposal of Solid Toxic Wastes*— The Report of the Technical Committee on the Disposal of Toxic Solid Wastes, Ministry of Housing and Local Government, Scottish Development Department, HMSO London, 1970.

3. 'The Analysis of Domestic Refuse', by A, E. Higginson. *A Review of Factors Affecting the Economic Collection of Refuse.* The Institute of Public Cleansing. (Now the Institute of Solid Wastes Management), 28 Portland Place, London W1N 4DE.

3a. *A Supplement to the Analysis of Domestic Refuse* by A. E. Higginson, The Institute of Public Cleansing.

4. *Notes on the Science and Practice of the Controlled Tipping of Refuse,* by R. E. Bevan, The Institute of Public Cleansing, London, 1967.

5. *Solid Wastes Disposal and Control,* Report of a WHO Expert Committee, WHO Technical Report Series No. 484, WHO, Geneva, 1971.

6. *Refuse Storage and Collection,* Report of the Working Party on Refuse Collection, Ministry of Housing and Local Government, HMSO, London, 1967.

7. *Industrial Wastes, A Working Party Document Embodying a Provisional Code of Practice for Disposal of Wastes,* The Institution of Chemical Engineers, 16 Belgrave Square, London SW1X 8PT. (1971).

## RECOMMENDED PRECAUTIONS FOR CONTROLLED TIPPING

1. Tipped refuse should be formed in layers not exceeding eight feet in depth after initial compaction.

2. The layer of refuse should be formed using a heavy tractor equipped with levelling blade. Refuse should be deposited on the tip surface and partially compacted by the tractor before being pushed over the face.

3. As tipping proceeds the tip faces and flanks should, wherever possible, be formed and consolidated to a gradient of about 1 in 3 by driving the tractor up and down the tip face.

4. The tipped maerial should be covered progressively so that all surfaces are covered with a layer of soil or other suitable inert material to a depth of about nine inches.

5. Large articles, such as furniture and hollow containers, should be broken up or crushed, deposited at the bottom of the tip face and covered by other refuse in such a position that they are not within three feet of the tip surface.

6. Any objectionable matter should be tipped at the bottom of the tip face so that it is immediately covered by other tipped material.

7. Suitable and adequate wire mesh screens should be erected in the vicinity of the tip face to obviate papers, plastic materials, etc., being blown away and deposited elsewhere on the tip surface.

8. The tip surface should be maintained in a tidy condition by the regular removal of all loose refuse.

9. No waste materials should be burned within the tip site and suitable emergency measures should be available for the prompt extinguishing of any fires which may occur.

10. Appropriate control measures against insect and rodent infestation should be undertaken on a routine basis.

11. In exposed locations or in the interest of visual amenity suitable screening by means of temporary fencing or other means should be provided around the tip site.

12. Suitable precautions should be taken in all cases to avoid pollution of surface or underground waters by percolate or leaching from the tip material.

## Air Navigation (Noise Certification and Limitation) Order 1972

I, BRIAN LENIHAN, Minister for Transport and Power, in exercise of the powers conferred on me by sections 5, 9, 11 and 16 of the Air Navigation and Transport Act, 1946 (No. 23 of 1946) (as amended by section 9 of the Air Navigation and Transport Act, 1950 (No. 4 of 1950)) and the Transport Fuel and Power (Transfer of Departmental Administration and Ministerial Functions) Order, 1959 (S.I. No. 125 of 1959), hereby order as follows:

Short title and commencement

1. (1) This Order may be cited as the Air Navigation (Noise Certification and Limitation) Order, 1972.

(2) This Order shall come into operation on the 12th day of December, 1972.

Definitions

2. In this Order:

'authorised person' means a person authorised under section 64 of the Air Navigation and Transport Act, 1936;

'by-pass ratio means the ratio of the air mass flow passing through the by-pass ducts of a gas turbine engine to the air mass flow passing through the combustion chambers of that engine, calculated at maximum thrust when the engine is stationary in an International Standard Atmosphere at sea level;

'Chicago Convention' has the meaning assigned to it by section 2 of the Air Navigation and Transport Act, 1946 (No. 23 of 1946), and includes any annex to the Chicago Convention relating to international standards and recommended practices and adopted in accordance with the Chicago Convention, and any amendment of any such annex made in accordance with the Chicago Convention;

'EPNdB' means a unit for expressing effective perceived noise level deermined in accordance with the method described in Appendix I of Annex 16 to the Chicago Convention;

'International Standard Atmosphere at sea level' means an atmosphere of which the characteristics are as follows:

(a) the air is a perfect dry gas;

(b) the physical constants are:

(i) sea level mean molecular weight:
$M_o + 28.9644 \times 10^{-3}$Kg $Mol^{-1}$,
(ii) sea level atmospheric pressure:
$P_o + 1013.250$ millibars,

(iii) sea level temperature: $t_o = 15°C$
$T_o = 288.15°K$,

(iv) sea level atmospheric density:
$p_o = 1.2250$ Kg $m^{-3}$,

(v) temperature of the ice point: $T_i = 273.15°K$,

(vi) universal gas constant: $R^* = 8.31432$ joules $(°K)^{-1}$ $mol^{-1}$;

'maximum total weight authorised' means the maximum total weight of the aircraft and its contents at which the aircraft may take off anywhere in the world in the most favourable circumstances in accordance with the certificate of airworthiness in force in respect of the aircraft;

'the Minister' means the Minister for Transport and Power;

'subsonic aeroplane' means an aeroplane incapable of sustaining level flight at speeds exceeding the speed of propagation of sound pressure waves in air.

Purpose of Order.

3. This Order is for the purposes of —

(i) giving effect to Annex 16 of the Chicago Convention in respect of noise certification of aeroplanes, and

(ii) making such directions as may be necessary from time to time in respect of procedures for the limitation of noise made by aircraft in certain locations and in certain circumstances.

Application of Order.

4. This Order applies to every subsonic aeroplane having turbine jet or turbine fan engines which—

(a) has a maximum total weight authorised of more than 5,700 Kg.,

(b) requires for take-off a runway length in excess of 450 metres, and

(c) (i) is powered by engines having a by-pass ratio of 2 or more and for which a certificate of airworthiness was first issued on or after the 1st day of March, 1972, or

(ii) is powered by other classes of engines, unless it is an aeroplane in respect of which the application for a certificate of airworthiness for the prototype of the aeroplane was accepted before the 1st day of January, 1969, or another equivalent procedure was effected by the certificating authorities of the state in which the aeroplane was manufactured before that date.

Requirement of a Noise Certificate.

5. Within the State, no person shall land or take-off, or cause to be landed or to be taken off, an aeroplane to which this Order applies unless —

(a) there is in force in respect of that aeroplane a certificate, in this Order called a Noise Certificate, issued by the Minister under Article 6 of this Order or (in the case of an aeroplane not registered in the State) issued by the competent authority of the state in which the aircraft is registered and acceptable to the Minister as having been issued in compliance with, or containing requirements substantially equivalent to, the requirements of the Schedule to this Order and

(b) any conditions subject to which the Noise Certificate was issued are complied with.

Issue of a Noise Certificate.

6. (1) The Minister may issue (or, in the case of a Noise Certificate issued by the competent authority of another State, may validate) a Noise Certificate in respect of any aeroplane to which this Order applies if he is satisfied that the noise made by the aeroplane is not greater than the noise levels specified in the Schedule to this Order.

(2) An applicant for the issue or validation of a Noise Certificate shall furnish such evidence and shall submit the said aeroplane to such tests (which may include tests in flight) as the Minister may require, for the purpose of proving compliance with the requirements of the Schedule to this Order.

(3) The Minister may issue every Noise Certificate, subject to—

(a) a condition stating the maximum total weights at which the aeroplane may take off or land when showing compliance with noise requirements, and

(b) such other conditions relating to noise made by the aeroplane as he may think fit.

(4) If the weights used in establishing compliance of aeroplane with the noise level requirements of the Schedule to this Order or equivalent noise level requirements are less than the maximum authorised weights for take off and landing established under the applicable airworthiness requirements, these lesser weights shall be furnished as operational limitations in the flight manual of the aeroplane concerned required by the Minister under the Air Navigation (Airworthiness of Aircraft) Order, 1964 (S.I. No. 141 of 1964).

(5) Subject to this Article and Article 8 of this Order, a Noise Certificate shall remain in force without limit of time.

(6) A Noise Certificate issued or validated under this Article shall cease to have effect if the aeroplane is modified in any way which affects its compliance with the noise standards required by this Order.

Document attesting noise certification.

7. (1) A Noise Certificate shall include information stating —

(a) the State in which the Aeroplane is registered and the national registration number or mark thereof;

(b) the manufacturer's serial number;

(c) the manufacturer's type and model designation;

(d) a statement of any modification to the aeroplane needed for the purpose of ensuring compliance with the applicable noise certification requirements;

(e) the maximum weight at which compliance with the applicable noise certification requirements has been demonstrated;

(f) any provisions prescribed by the Minister in respect of different classes of aeroplanes and with respect to different circumstances.

(2) A Noise Certificate shall be carried in the aeroplane to which it applies, save that if a flight by that aeroplane is intended to begin and to end at the same aerodrome without an intermediate landing at another aerodrome, the Noise Certificate may be kept at the first-mentioned aerodrome.

(3) The pilot-in-command of an aeroplane, on being required by an authorised person to produce or cause to be produced for inspection by the authorised person the Noise Certificate in force in respect of the said aeroplane, shall produce or cause to be produced the said Noise Certificate.

Suspension, revocation or variation of noise certificates.

8. (1) The Minster may suspend any Noise Certificate, validation, approval, or other document issued under this Order and, after an investigation at which the person to whom the document has been issued has been given an opportunity of being heard, may revoke, suspend or vary any such document and in such a case the holder or any person having possession or custody of the document shall surrender it to the Minister within a reasonable time after being required to do so by him.

(2) The breach of any condition subject to which any Noise Certificate, validation, approval or other document has been issued under this Order shall render the document invalid during the continuance of the breach.

Fraudulent use or forgery of Noise Certificates.

9. A person shall not —

(a) forge or fraudulently alter, or procure to be forged or fraudulently altered or assist in forging or fraudulently altering, a Noise Certificate,

(b) use any Noise Certificate required under this Order which has been forged, fraudulently altered, revoked or suspended or has ceased to be in force, or

(c) give or lend any Noise Certificate to any other person or allow or permit it to be used by any other person.

False representations.

10. A person shall not make, procure to be made or assist in making, any false representation for the purpose of procuring for himself or for any other person the issue, validation, renewal or variation of a Noise Certificate.

Right of access to and detention of aeroplanes.

11. (1) Whenever it appears to the Minister or an authorised person that an aeroplane is intended or likely to be flown from any place in the State in such circumstances that the flight would be a contravention of Article 5 of this Order or of any other provision of this Order or any direction made thereunder, the Minister or authorised person may give to the operator of the aeroplane or to the pilot-in-command thereof such instructions, and take such steps for the detention of the aeroplane or otherwise, as appear to him to be necessary in order to prevent the flight, and a person so instructed shall comply with the instructions.

(2) For the purposes of paragraph (1) of this Article the Minister or authorised person concerned may enter upon and inspect the aeroplane concerned.

Right of access to aerodrome and other places.

12. The Minister or an authorised person may, for the purpose of ascertaining compliance with the provisions of this Order, have access at all reasonable times to any aerodrome or place within the State for the purpose of inspecting either any aeroplane on the said aerodrome or place or any document which under this Order he has power to demand, or for the purpose of detaining any

aeroplane under this Order, but access to an aerodrome under the control of the Minister for Defence shall be obtained only with the permission of the person in charge of the aerodrome.

Obstruction of authorised persons.

13. A person shall not wilfully obstruct or impede an authorised person acting in the exercise of his powers or the performance of his duties under this Order.

Enforcement of directions.

14. Any person who fails to comply with a direction given to him by the Minister or an authorised person under any provision of this Order shall be deemed for the purpose of this Order to have contravened that provision.

Exemptions

15. The Minister may exempt from any of the provisions of this Order any classes of aeroplanes.

Directions under this order.

16. (1) The Minister may give directions for carrying out the purposes of this Order.

(2) Directions under this Order may be given in the form of Notices to Airmen (otherwise known as Notams), Notices to Aircraft Owners and Aircraft Engineers, Aeronautical Information Circulars or by notices sent by registered post to the person affected.

## SCHEDULE

**Article 6**

**Requirements for the issue of a Noise Certificate**

1. In this Schedule —

'the noise certification reference conditions' means conditions in which—

(i) the sea level atmospheric pressure is 1013.25 millibars;

(ii) the ambient air temperature is 25° Celsius;

(iii) the wind velocity is zero;

(iv) the relative humidity is 70% and

(v) the maximum take-off and landing weights of the aeroplane are those at which noise certification is requested by the applicant for a Noise Certificate.

2. The noise evaluation measure shall be the effective perceived noise level expressed in units of EPNdB as described in Appendix 1 of Annex 16 to the Chicago Convention.

3. Subject to the provisions of paragraph 6 of this Schedule an aeroplane shall not, when flying in the noise certification reference conditions set out in paragraph 1 of this Schedulc, exceed maximum noise levels at —

(a) lateral and approach noise measuring points located as described in sub-paragraph (1) and sub-paragraph (2) of paragraph 4 of this Schedule of 108 EPNdB for aeroplanes with a maximum total weight of

272,000 kg. or more, less 2 ENPdB per halving of the 272,000 kg. maximum weight, down to 102 ENPdB for maximum total weights authorised of 34,000 kg. or less, and at

(b) a flyover measuring point located as described in paragraph 4 of 108 EPNdB for aeroplanes with a maximum total weight authorised of 272,000 kg. or more less 5 EPNdB per having of the 272,000 kg. maximum weight own to 93 EPNdB for maximum total weights authorised of 34,000 kg. of less.

4. (1) The lateral noise measuring point is a point on a line parallel to and 650 metres from the centre line, or extended centre line of the runway where the noise level is greatest during take off.

(2) The approach noise measuring point is a point on the extended centre line of the runway 120 metres vertically below a 3 degrees descent path originating from a point 300 metres along the runway from the threshhold.

(3) The flyover noise measuring point is a point on the extended centre line of the runway at a distance of 6,500 metres from the start of the run to take-off.

5. The values of noise levels specified in sub-paragraph (a) and sub-paragraph (b) of paragraph 3 of this Schedule may be required to be determined by air test of the aeroplane concerned in accordance with conditions which are —

(i) in compliance with the specifications in this Schedule and

(ii) in compliance with the requirements and methods set out in Annex 16 to the Chicago Convention.

6. The noise levels specified in paragraph 3 of this Schedule may be exceeded at one or two of the measuring points specified in paragraph 4 of this Schedule if —

(a) the sum of the excesses is not greater than 4 EPNdB, except that in respect of four-engined aeroplanes powered by engines with a by-pass ratio of 2 or more and for which the application for a certificate of airworthiness for the prototype of the said aeroplane accepted or another equivalent procedure was effected by the certificating authorities of the state in which the aeroplane was manufactured, prior to the 1st of December 1969, the sum of any excesses shall not be greater than 5 EPNdB;

(b) at no measuring point is the excess more than 3 EPNdB and

(c) the excesses are completely offset by reductions at the other measuring points.

**Explanatory Note**

(This note is not part of the Order and does not purport to be an interpretation of it)

1. This Order applies to every subsonic aeroplane having turbine jet or turbine fan engines which

(a) has a maximum total weight authorised of more than 5,700 kilograms;

(b) requires for take-off a runway length in excess of 450 metres; and

(c) (i) is powered by engines having a by-pass ratio of 2 or more and for which a certificate of airworthiness was first issued on or after 1st March, 1972, or

(ii) is powered by other classes of engines unless it is an aeroplane in respect of which the application for certificate of airworthiness for the prototype of the aeroplane was accepted before 1st January, 1969, or another equivalent procedure was effected by the certificating authorities of the state in which the aeroplane was manufactured before that date.

2. It prohibits an aeroplane to which it applies, wherever registered, from landing or taking off in the State except in accordance with a noise certificate issued by the Minister or the competent authority of the State of registration and acceptable to the Minister.

3. It makes provision for the issue of noise certificates by the Minister, specifies standards with which an aeroplane must comply in order to obtain such a certificate and contains ancillary provisions including power to detain an aeroplane in order to prevent a contravention of the Order.

4. It gives effect to the standards in Annex 16 of the Chicago Convention.

## Air Navigation (Noise Certification and Limitation) (Amendment) Order, 1973

I, PETER BARRY, Minister for Transport and Power, in exercise of the powers conferred on me by sections 5, 9, 11 and 16 of the Air Navigation and Transport Act, 1946 (No. 23 of 1946) (as amended by section 9 of the Air Navigation and Transport Act, 1950 (No. 4 of 1950)), and the Transport, Fuel and Power (Transfer of Departmental Administration and Ministerial Functions) Order, 1959 (S.I. No. 125 of 1959), hereby order as follows:

1. (1) This Order may be cited as the Air Navigation (Noise Certification and Limitation) (Amendment) Order, 1973.

(2) The Air Navigation (Noise Certification and Limitation) Order, 1972 (S.I. No. 280 of 1972), and this Order shall be construed together and may be cited together as the Air Navigation (Noise Certification and Limitation) Orders, 1972 and 1973.

2. The Air Navigation (Noise Certification and Limitation) Order, 1972, is hereby amended by the insertion after Article 4 of the following Article:

2. The Air Navigation (Noise Certification and Limitation) Order, 1872, is hereby amended by the insertion after Article 4 of the following Article:

4. This Order also applies to every subsonic aeroplane having turbine jet or turbine fan engines—

(a) which has a maximum total weight authorised of over 28,500 Kg.,
(b) which is powered by engines having a by-pass ratio of less than 2,

(c) for which the application for a certificate of airworthiness for the prototype of the aeroplane was accepted, or another equivalent procedure was effected by the certificating authorities of the state in which the aeroplane was manufactured, before the 1st day of January 1969, and

(d) for which a certificate of airworthiness for the individual aeroplane is first issued on or after the 1stday of January 1976.'

**EXPLANATORY NOTE**
(This note is not part of the Instrument and does not purport to be a legal interpretation).

The purpose of this Order is to extend, from 1 January, 1976, to certain larger type subsonic aircraft previously exempted, the requirements of the Air Navigation (Noise Certification and Limitation) Order, 1972 regarding noise levels.

It gives effect to the standards in Amendment Number 1 to Annex 16 to the Chicago Convention.

# SOUTH-WEST REGION CORK-KERRY OIL POLLUTION CLEARANCE SCHEME OPERATIONAL DIRECTORY

(1st Edition)

**SECTION 1**
**Regional Coordination**

| *Designation* | *Name* | *Tel. No.* | *After Hours Tel. No.* |
|---|---|---|---|
| 1. Regional Coordinator | M. Whelan | (021) 26891 | |
| 2. Deouty Regional Coordinator | | | |
| 3. Regional Administrative Back-up (Regional Development Organisation) | J. A. Quish | (021) 25337/8 | 44173 |

**SECTION II**
**Cork County Council**

Tel. (021) 26891

| *Designation* | *Name* | *Tel. No.* |
|---|---|---|
| 1. Oil Pollution Officer | M. Whelan | (021) 26891 |
| 2. Deputy Oil Pollution Officer | | |
| 3. Planning Officer | F. Jordan | |
| 4. County Engineer | D. Kavanagh | (021) 26891 |
| 5. County Manager | M. N. Conlon | (021) 26891 |
| 6. County Secretary | J. K. O'Hare | (021) 26891 |
| 7. County Solicitor | J. Hickey | (021) 26891 |
| 8. C.A.C.E. West Cork | M. W. Mullins | Skibbereen 3 |
| 9. C.A.C.E. South Cork | M. Whelan | (021) 26891 |
| 10. A.C.E.'s West Cork: | | |
| Bantry | J. F. Matson | Bantry 48 |
| Castletownbere | H. B. Roche | Castletown 17 |
| Clonakilty | J. J. McCarry | (023) 43347 |
| Schull | M. Curtain | Schull 10 |
| Skibbereen | P. J. Daly | Skibbereen 3 |
| 11. A.C.E.'s South Cork: | | |
| Youghal | F. X. Henegan | (024) 2365 |
| Crosshaven | D. Dunnion | (021) 26891 |
| Kinsale | P. J. Treacy | (021) 72137 |

**Equipment and detergent available**

**a. Equipment**

| *Name* | *Type* |
|---|---|
| Spray Equipment | 30 Knapsack Sprays<br>Traxcavators (etc.) and Trawlers at Castletownbere, Schull, Castletownsland, Glanmore to be mobilised. |

**b. Detergent**

| *Type Name/No.* | *Amount lb/gal.* |
|---|---|
| Permission sought to buy 3,000 gal. | |

**SECTION III**
**Cork Harbour Commissioners**

| *Designation* | *Name* | *Tel. No.* | *After Hours Tel. No.* |
|---|---|---|---|
| 1. Oil Pollution Officer | Capt. P. Tyrrell | (021) 23125 | (021) 821231 |
| 2. Deputy Oil Pollution Officer | P. Mullan | (021) 23125 | (021) 811688 |
| 3. General Manager | L. French | (021) 23125 | |
| 4. Chief Engineer | H. O'Neill | (021) 23125 | |
| 5. Harbour Master | Capt. P. Tyrrell | (021) 23125 | (021) 821231 |
| Cork Corporation | | | |
| 1. Oil Pollution Officer | J. J. Cronin | (021) 21731, ext. 63 | (021) 31450 |
| 2. Deputy Oil Pollution Officer | E. Hayes | (021) 21731 | (021) 43254 |
| 3. City Manager | P. Clayton | (021) 21731 | |
| 4. City Engineer | S. McCarthy | (021) 21731 | |
| 5. C.A.C.E. | E. P. D'Alton | (021) 21731 | (021) 33495 |
| Marine Transport Services Ltd., Atlantic Quay, Cobh | B. Foley | (021) 811485 | (021) 811414 |
| | E. Perry | (021) 811485 | (021) 811223 |

**Equipment and detergent available**

**a. Equipment (craft)**

| *Name* | *Type* | *O.A.L.* | *Capability* | *Availability* |
|---|---|---|---|---|
| Knapsack Sprayers | Spray Equipment | | | |
| Grabwell | Dredger | 115 feet | Harbour and Approaches | Only in serious emergency and limited periods |
| Richard Wallace | Tug | 71 feet | ditto | ditto |
| Shandon | Tug | 87 feet | ditto | ditto |
| Carraig An Cuain | Motor Launch | 35 feet | Harbour | ditto |
| St. Nessan | Pilot Launch | 45 feet | Harbour and Approaches | ditto |
| St. Colman | Pilot Launch | 37 feet | ditto | ditto |

**b. Detergent**

| *Type Name/No.* | *Amount lb./gal.* |
|---|---|
| Correxet | 70 gallons |

**Cork Corporation**

**Equipment and detergent available**

**a. Equipment**

| *Name* | *Type* |
|---|---|
| Spray Equipment | Knapsack unit to be available on loan from ports department pending acquisition of own set<br>2 No Oil bantex flock to absorb small oil spills |

| | |
|---|---|
| Boats | Hire arrangements made with Cork Hrabour Commissioners |
| Other Equipment | NIL Hire J.C.B. machines, could be diverted from other work at short notice in case ofemergency |

**b. Detergent**

| *Type Name/No.* | *Amount* |
|---|---|
| Slicgol 505 | 45 gal. or 200 litres |
| Gambin 105 | 45 gal. or 200 litres |

**SECTION IV**
**Kerry County Council**

| *Designation* | *Name* | *Tel. No.* | *After Hours Tel. No.* |
|---|---|---|---|
| 1. Oil Pollution Officer | A. N. Dillon | (066) 21111 | (066) 21220 |
| 2. Deputy Oil Pollution Officer | D. J. O'Callaghan | (066) 21111 | (064) 31152 |
| 3. County Manager | A. Stack | (066) 21111 | |
| 4. Assistant County Manager | | (066) 21111 | |
| 5. County Engineer | P. J. Malone | (066) 21111 | |

**Equipment and detergent available**

**a. Equipment**

| *Name* | *Type* |
|---|---|
| Spray Equipment | 2 Warren Spring Mini Packs |
| Boat Equipment | Boat available at Fenit, Dingle, Caherciveen |
| Other equipment | Adequate number available |

**b. Detergent**

| *Type Name/No.* | *Amount* |
|---|---|
| BP 1100X | 90 by 45 gal. drums |

**SECTION V**

**Government Departments**

| *Designation* | *Name* | *Tel. No.* |
|---|---|---|
| Local Government Environment: | | |
| Principal Officer | J. McGrath | (01) 42961 |
| | B. Cambell | (01) 42961 |
| | M. Phelan | (01) 42961 |
| Department of Defence: | | |
| Haulbowline | | (021) 811246 |
| | | (021) 841353/4 |
| Helicopter Service Balldonnell | | (01) 21202 |

| | | |
|---|---|---|
| Department of Transport and Power | | (061) 61969 |
| Marine Rescue Coordinating Centre | Shannon | (061) 61219 |
| | Airport | (021) 26552 |
| Meteorological Service Cork Airport | C. O'Connor | (021) 25341 |

**SECTION VI**

**Other Harbour Masters**
Bantry
Castletown Bere
Schull
Skibbereen
Youghal
Crosshaven
Kinsale

**SECTION VII**

**Other Concerned Organisations**

| *Designation* | *Name* | *Tel. No.* | *After Hours Tel. No.* |
|---|---|---|---|
| Gulf Oil Terminals | A. J. Monaghan | Bantry 381 | Durrs 55 |
| | E. Beyer | Bantry 282 | Durrs 44 |
| | Supervisors: | Bantry 380 | |
| | L. K. Heppes | | |
| | W. K. Ellis | | |
| Marathon Petroleum Ireland Ltd., | G. Susich | (021) 42887 | (021) 871248 |
| | R. L. Trimbell | (01) 781955 | (01) 986555 |
| Irish Refining Co. Ltd. | L. McGovern | (021) 25144, ext. 320 | (021) 63396 |
| | J. Lyall | | |
| E.S.B. | D. Clarke | (021) 24251 | |
| Irish Shell & B.P. Ltd. | | | |
| Esso (Theoranta) Ltd., | | (021) 25351 | |

**Marine Transport Services Limited**
**Equipment Available (Craft)**

| *Name* | *Type* | *O.A.L.* | *Capability* | *Availability* |
|---|---|---|---|---|
| Spray Equipment | One pair spray booms approx. 12ft long with pump, 4 nozzles and gate type agitator, for fitting to tug *Anarica*. (This equipment is not to W.S. design but is somewhat similar.) | | | |
| Anarica | Tug | 87 feet | Harbour and Approaches | At any time |
| William J | Tug | 81 feet | ditto | ditto |
| NGOT | Tug | 65 feet | ditto | ditto |
| John L | Tug | 62 feet | ditto | ditto |
| Marian | Launch/Workboat | 48 feet | ditto | ditto |
| "190" | ditto | 50 feet | ditto | ditto |
| Greata Ban | ditto | 47 feet | ditto | ditto |
| Thomas O.S. | ditto | 48 feet | ditto | ditto |
| Billet | ditto | 32 feet | ditto | ditto |
| Pullit | ditto | 32 feet | ditto | ditto |
| *Celt | Self propelled Lighter | 80 feet | ditto | ditto |
| *Corpach | ditto | 80 feet | ditto | ditto |
| Miller | ditto | 85 feet | ditto | ditto |
| Patrina | Launch/Workboat | 50 feet | ditto | ditto |
| Regina | ditto | 40 feet | ditto | ditto |
| Eileen | ditto | 30 feet | ditto | ditto |
| Failte | ditto | 27 feet | ditto | ditto |

In addition to above 2 dumb barges capable of carrying up to 200 tons and 1 dumb barge capable of carrying 40 tons.

**Fitted with winch and derrick.*

**Gulf Oil Terminals (Ireland) Ltd.**
**a. Equipment available**

| *Name* | *Type* |
|---|---|
| Spray equipment | 6 Knapsack type sprayers. All boats fitted with spray lance equipment and spray pumps |
| Boats | 4 x 2500 hp Tugs,<br>2 x 40 feet 160 hp workboats.<br>1 x twin 360 hp foot personnel/pilot boat |
| Other equipment | 1000 metres T.T. Oil boom in length of 165 feet (50 metres)<br>1 x 30 cubic metre oil skimmer barge with meter driven pumps |
| On order: | One Warren Springs Inshore Anti-Pollution Unit.<br>Onc "Manta Ray" skimmerwith portable pump. This will be used when necessary form a line boat and any oil recurred will be pumped into a floating 5000 gallon container now being adapted for marine use. |

**b. Detergents available**

| *Type Name/No.* | *Amount* |
|---|---|
| Jasol No. 5 | 20,000 Gallons |

(Note: Jasol will be no longer purchased and approved type only purchases)
(Note: A quantity of moss peat is also maintained at Terminal)

**Marathon Petroleum (Ireland) Ltd.,**
**Equipment available**

No oil pollution control equipment available at the present time. When their operations are active, they will attempt to keep on hand in Cork approximately fifty 45 gallons each drums of BP 100X.

They would expect to obtain additional equipment if they are successful in their offshore exploration activities.

**Irish Refining Co. Ltd.,**

**a. Detergent available**

| *Type Name/No.* | *Amount lb./gal.* |
|---|---|
| B.P. 1100 | 700–2,500 gal. |

**b. Equipment available**

| *Name* | *Type* |
|---|---|
| Spray equipment | (a) Hathaway Portable Pumps 7 g.p.m. which can be used from launch or shore<br>(b) 5 Portable knapsack sprayers |
| Boats | 2 x 17 g.p.m. pumps, one on each of the refinery fine tugs and associate equipment as specified by Warren Springs Lab. to D.T.I. |
| Other equipment | 1 Tractor and scoop shovel<br>1 Truck, equipped with crane, available locally straw, bulldozer, launches |

**Various fishing boats whose owners have indicated willingness to work on oil pollution**
**Equipment available**

| *Name of Boat* | *O.A.L.* | *Capability* | *Owner and Address* | *Tel. No.* |
|---|---|---|---|---|
| Dinausaure | 40 feet | Harbour and Approaches | R. C. Thompson, 1 Bellevue, Cobh | (021) 811616 |
| Inin Mor | 36 feet | ditto | W. Frahill, Cor, Maria, Cobh. | (021) 811709 |
| Lis Mor | 30 feet | ditto | M. Taft, 22 Grahams Terrace, Cobh | (021) 811711 |
| Dun Laoire | 45 feet | ditto | A. J. Stoat, 7 West View, Cobh | (021) 811283 |
| Clonlara | 26 feet | Harbour | J. Cody, 22 Calvins Terrace, Cobh | None |
| Uam Var | 26 feet | ditto | K. Jones & J. Stoat, 9 West View, Cobh | Via (021) 811283 |

## SECTION VIII

**Scientific advisers**

| *Designation* | *Name* | *Tel. No.* |
|---|---|---|
| 1. Dept. of Local Government | J. J. Coffey, A.C.E.A. Environment | (01) 42961 |
| 2. Dept. of Agriculture & Fisheries | D. De Griffith, | (01) 67561 |
| | P. Maloney | (01) 41121 |
| 3. Institute of Industrial Research and Standards: | T. McManus | (01) 370101 ext. 366 |
| Dublin | D. M. Carroll | (01) 37101 ext. 329 |
| Cork | S. O'Neill | (021) 24341 |
| 4. Cranford Technical Institute, Dept. Chem. Technology | | |
| 5. University College, Cork (Department of Chemistry) | Prof. F. L. Scott | (021) 26871 |
| 6. Bord Na Mona | F. Lunny | (01) 66431<br>(045) 31201 |
| 7. Warren Spring Laboratory | Dr. P. G. Jeffery | |

## SECTION IX

**Miscellaneous**

| *Designation* | *Name* | *Tel. No.* |
|---|---|---|
| Regional Coordinator, Mid-West Region | | |
| Regional Coordinator, South East Region | | |
| Oil Pollution Officer, Co. Waterford | | |
| Oil Pollution Officer, Co. Limerick | | |
| Meteorological Office, Cork Airport | C. O'Connor | (021) 26786 |
| Valencia | S. McWilliams | Caherciveen 27 |
| Dublin | F. E. Dixon | (01) 43320<br>(01) 778251<br>(01) 44433 |
| Marine Rescue Coordinating Centre, Shannon Airport | | (061) 61969<br>(061) 61219 |
| Department of Defence, Dublin | S. L. O'Riain, Capt. T. McKenna | (01) 771881 |

**Limerick, Clare and Tipperary (N.R.) Region and North Kerry**
**Anti-Oil Pollution Plan**
for area from Tralee Bay to Galway Bay including the Shannon Estuary.

Should oil pollution occur in the area this plan sets out:

1. The reporting procedure
2. The resources available
3. The personnel available
4. The agreed action to be taken

**1. Reporting of oil pollution**

The waters of entire Shannon Estuary, Tralee Bay and part of Galway Bay form parts of harbours. The Oil Polltion of the Sea Act 1956 requires oil pollution within a harbour to be reported forthwith to the harbour master. Such reports will be transmitted to the local oil pollution officer and the coordinator.

Shipmasters, ferrymasters, gardai, fishermen, lighthousekeepers, as well as government, local authority and harbour personnel shall be made aware of the importance of promptly reporting any oil pollution. They must have readily available the names and telephone numbers of the local pollution officers and their deputies.

A circular setting out the reporting procedure (see appendix) shall be romulgated to:

1. All key local authority personnel
2. All harbour and pier personnel including pilots
3. Local ships' agents
4. E.S.B. Fisheries division
5. Skippers of fishing boats and ferries within the region.
6. Inshore fishermen (through Fishery Conservators etc.)
7. All garda stations on or near the coastal perimeter or the region
8. Full-time and part-time Lighthouse Keepers in the area
9. Defence forces, including Civil Defence
10. Sporting organisations including Yacht Clubs, Fishing Clubs, Golf Clubs on the coast etc.
11. Local Conservation organisations i.e. An Taisce, Limerick Wildlife ISPCA etc.
12. Operators of industries etc. sited close to the shore line
13. Marine Rescue Coordination Centre, Shannon
14. Shannon Airport Rescue Service
15. Local branches of IFA, Macra na Feirme, etc.
16. All newspapers circulating in the area

**RESOURCES AVAILABLE**

**Clare County Council**

*At Ennis:*

2025 gal. BP 1100X dispersant

1 mechanical dispersant pump, single lance, plus 300 feet of hose.

Protective clothing

*At Kilrush*

2025 gal. BP 1100X dispersant

protective clothing

*Note:* 2 'Cooper Peglar' oil dispersant demountable trailer pump units are due for deliver in April 1974

| *General equipment* | *Council owned* | *On hire* |
|---|---|---|
| Mechanical shovels | 3 rubbers, 5 tracked | 8 |
| Excavators | 7 | — |
| Trucks | 23 nos. 3-16 cu. yd. | 6 x 8 cu. yd. licensed<br>20 x 6-8 cu. yd. unlicensed |
| Bulldozers | — | 5 |
| *Sprayers | 5 plus 1 trailer | — |
| Tractors and trailers | — | many readily available |
| Fire Brigade | 5 large appliances (500 gpm)<br>5 portable pumps (250 gpm) | |

*V.H.F. radio chain:* Base station at Ennis with 11 vehicles mounted mobiles of which 5-6 would be available for oil pollution.

*Boats:* GalwayBay –12 trawlers 50-75 feet several lobster boats.

'Tug-tender'–GalwayBay available part of the year.

**Sprayers would require twenty four hours cleaning out.*

**Kerry County Council**

*At Tralee:*

4050 gals BP 1100X

2 Warren Springs inshore minipack units

Knapsack sprayers

50 sets gloves, goggles, boots.

| *General equipment* | *Council owned* | *For hire* |
|---|---|---|
| Mechanical Shovels (Rubbers) | 3 | — |
| Traxcavators | 6 | 6 |
| Trucks | 20 | 15 |
| Bulldozers | — | 10 |
| Excavators | — | 6 |
| Trenchers | — | 18 |

**Fire Brigade equipment**

*At Tralee:*

1 Dennis 950 gpm engine

1 'Coventry Climax' portable pump 250 gpm

1 Civil Defence fire engine (green goders)

1 Civil Defence 'Goodwin' portable pump

*At Listowel:*

1 'Bedford' appliance 650 gpm

1 'Godiva' portable pump 350 gpm.
*Boats:* Trawlers for hire at Fenit 2 x 70 feet, 3 x 50 feet, 1 x 40 feet.
*V.H.F.Radio:* None at present but some facilities expected by end of 1974.

**Limerick County Council**
**Special equipment**
*At Newcastle West:*
315 gal. BP 1100X dispersant
12 Knapsack sprays
12 each jackets, pants, boots and goggles
1 cwt. barrier cream

| *General equipment* | *Council* | *On offer for hire* |
|---|---|---|
| Excavators | — | 25 |
| Traxcavators | 1 | 16 |
| Bulldozers | — | 20 |
| Loaders | 2 | 9 |
| Dumpers | 2 | 20 |
| Lorries | 26 | 21 |
| Tractors | — | 3 |
| Pickups | 7 | — |
| Vans | 14 | — |

*VHF Radio* not installed at present

**Limerick Corporation**
*Special equipment*
At Limerick: 3 Knapsack sprayers
*General equipment:*
5 domestic refuse side loaders
7 pick-up trucks
1 van
2 tractors (1 with trailer)
4 ambulances
4 fire engines
1 Landrove
1 rescue truck

**Limerick Harbour Commissioners**
*At Limerick Dock:*
1 x 2 ton tipper truck
100 gals 'Polyclens' detergent
100 gals. 'Jaysol' detergent
300 gal. 'BP 1100X' dispersant
Knapsack sprayers
1 x 150 feet oil boom
¾ ton milled peat in bales

*Vessels:*
1 Grab/suction dredger *Curraghgour II* 140 feet 520 bhp normally based Limerick.
2. *Shannon Pilot* 51 feet pilot launch, based Cappa, Kilrush
3. *Patricia* 40 feet ex-pilot boat, based Limerick
4. *St.Munchin* 36 feet workboat, based Limerick
*VHF Radio:* 2 x 25 Mar VHF Bases Station (Limerick Docks and Cappa Kilrush)
2 Mobile 10w and 2 mobile 1w *Curraghgoer II* and *Shannon Pilot* are VHF equipped.

**Foynes Harbour Trustees**
1 portable 'Spate' pump 250 gpm
1 x 10 ton tipper truck
21 feet workboat
*VHF Radio:* 1 x 25 w Marine VHF Base Station

**Cement Ltd.**
*At Mungret, Limerick:*
45 gal. 'Jaysol' Detergent
*At Foynes:*
225 gal. 'Jaysol' detergent
*At Foynes Island:*
200 feet x 6 inch Slick-bar oil boom
*Vessels:*
1 catamaran landing craft (will carry one truck or bulldozer)
1 x 27 feet workbot (both based Foynes)

**ESB Tarbert**
*At Tarbert*
500 gal. BP 1100X dispersant
2 x 200 feet 6 inch by 6 inch Slickbar oil booms
3 x 50 feet absorbent oil booms
1000 feet ½ inch dia. rope, 10 x 42 inch mooring buoys and six anchors for booms)
4 Knapsack sprayers and 1 motorised sprayer
1 Barrel bung sprayer
12 bales straw
20 cwt. milled peat
100 plastic bags, 288 sandbags
30 feet workboat at Ballylongford, 16 feet dinghy with 9 hp outboard at station
*Communication*
Station connected to internal ESB phone system. Could be useful in emergency if P & T system suffers breakdown. Some 'walkie-talkies'.

**Shannon Meteorological Station**
Weather information service

**Marine Rescue Coordinator Centre Shannon**
Communication facilities. Normal function is coordination of land, sea and air facilities for rescue operations at sea.

**Shannon Airport**
*At Dernish Island:*
2 rescue launches, 1 amphibious vehicle on 24 hour stand-by for rescue operations

**Conservation Societies**
Limerick/Clare Branch Irish Society for Prevention of Cruelty to Animals: Facilities to deal with oil affected birds, etc.
*Conservation of wild life:* Information and contacts

**PERSONNEL AVAILABLE**
**Clare County Council**

Pollution Officer: Mr. E. Duffy, Ennis. Tel. office 21616, home 216254
In Mr. Duffy's absence: Mr. Piggot, Ennis. Tel. office 2616, home 21080
Mr. Torpey, Ennis. Tel. office 21616, home 21507
Manpower: 500 County Council employees
100 Civil Defence personnel
30 Fire Service personnel
*Note:* The inner part of Galway Bay comprises the Port of Galway and the office with responsibility under the 1956 Pollution of the Sea Act is: Captain J. Whyte, Harbour Master. Tel. office (091) 2329, home (091) 2758

**Kerry County Council**
Oil Pollution Officer: Mr.Dillon,Tralee.Tel. office (066 21111, home (066) 21220
Mr.D. J. O'Callaghan, Tralee. Tel office (066) 21111
Useful contact: Mr. 'ohn McCarthy, Machinery Overseer, Tel. home (066) 22381
*Tralee Bay Area:*
Office with special responsibility for oil pollution under 1956 Oil Pollution of the Sea Act. Cdr. J. Barrett, Harbour Master Fenit, (066) 36103.
Manpower: 361 road workers
30 Civil Defence
12 Fire Brigade, Tralee
12 Fire Brigade, Listowel

**Limerick County Council**
Oil Pollution Officer: Mr. P. King. Tel. Rathkeale 10 (office). Lives Foynes, nearby phone Foynes 40 or Foynes Gardai.
Mr. N. Hogan (Adare) Tel. 94228
Overseer at Newcastle West is Mr. M. O'Connor
Manpower: Rathkeale area 60 men

**Limerick Corporation**
Oil Pollution Officer Mr. G. Tomany Tel. office (061) 45799 home (061) 44703
Manpower: 20 men (including supervisors, gangers and drivers but not tradesmen)

**Limerick Harbour Commissioners**
Oil Pollution Officer and Regional Coordinator: Capt. E. K. Donnelly, Harbour Master. Tel. office (061) 44914, home (061) 48862
Deputy Pollution Officer: Capt. D. J. McLaughlin, Assistant Harbour Master. Tel. office (061) 44914, home (061) 48474
Other useful numbers:
Harbour Engineer (061) 45109
Limerick Dock Main Gate (061) 45916
Dernish Jetty, Shannon (061) 61081
Manpower: 30 (including crews and watchkeepers)

**Foynes Harbour Trustees**
Oil Pollution Officer Deputy Regional Coordinator: Capt. J. O'Connor, Harbour Master. Tel. Foynes 71 (extension to home)
Manpower: 5 men.

**Cement Ltd.**
*Limerick:* Mr. P. Gleeson, Works Manager, office 44611
*Foynes:* Mr. M. Finnucane, Depot Foynes 37, home Foynes 73
*Foynes Island Jetty* (when vessels are discharging) Foynes 55

**ESB Tarbert**
Station Superintendent: Mr. Beary, Tarbert 64/65
Shift Supervisor: Mr. George Eacrett at station or home Tarbert 61
Manpower: 10-15 men and fitter and electrician

**Shannon Meteorological Station**
Supervisor (061) 61439
Forecaster (061) 61333 and (061) 61242

**Marine Rescue Coordinator Centre Shannon**

Duty Officer (061) 61219, (061) 61969
Duty Officer, Shannon Rescue (061) 61435

**Shannon Airport**
Duty Officer, Airport Engineer's Office, (061) 61222
Duty Officer, Control Tower (061) 61111

**Conservation Societies**
ISPCA Inspector Maloney (resident Limerick). Tel. (061) 45618
Limerick Wild Life Society: Mr. M. A. Quillinen. Tel. (061) 46895

**Other Concerned Organisations**
Esso Teoranta Limerick Depot: (061) 46388. Shannon Depot (061) 61217, Mr. McDonald (061) 61609
Irish Shell/BP: Limerick Depot: (061) 46222, Mr. Coughlan home (061) 46217
Texaco (Ireland) Ltd.: Limerick Depot (061) 45922, Mr. Perkins home (061) 49564

## AGREED ACTION TO BE TAKEN

1. **Inform County Oil Pollution Officer and Regional Coordinator, also local Harbour Master (if applicable).** This should be done on the line suggested in Appendix 1. (Reporting of oil pollution). The importance of educating every potential observer to report in this form cannot be overemphasised. In order to act promptly, decisively and to best effect the officer concerned must have as much information as possible as early as possible. The Regional Coordinator will inform the National Coordinator.

2. **Alert key personnel.** This would normally be the local county personnel, but if the neighbouring county is also threatened they would be alerted through their own oil oil pollution officer. Each county would have its own plan for 'alert'.

3. **Identify the problem.** Best use should be made at this stage of the Meteorological Service, tidal information and corrobative local information to identify danger areas and to obtain the best estimate of the position and extent of the slick.

4. **Call out key personnel.** as indicated above (including boat skippers if necessary).

5. **Shift equipment to danger/polluted areas** as indicated above. Boats to load at designated berths. Take samples before treatment to assist in tracing culprits.

6. **Set up communications headquarters.** To direct, coordinate and supply anti-pollution activities, communication must be established between units at sea, on beaches and sources of manpower and equipment. Another important function would be to keep the news media properly informed of the situation as it develops as well I.SPCA etc. The National Coordinator, the Departments of Local Government, Agriculture and Fisheries, Transport and Power, etc. should be kept abreast of the situation.

In general two communication centres will be established

a. Regional Headquarters which will normally be set up at the Harbour Master's Office in Limerick and

b. Local headquarters at or near the site.

7. **Call in resources from adjacent authorities** and back-up services as anti-pollution requirements dictate. There sources of each authority and organisation concerned will be made available if required to each other in dealing with oil spillage either offshore or on shore.

8. **Hold daily briefing sessions** preferably at the local headquarters in communication with the Regional Headquarters, so that all officers engaged in the anti-pollution activities will be fully informed of developments and plans may be drawn up to deal with operational difficulties encountered such as disposal of cleared oil, etc.

9. **Hold press conferences.** At the conclusion of each daily session, press representatives, conservancy organisations etc. will be briefed by either the county pollution officer or the regional coordinator.

10. **Keep records of expenditure:** At all times sufficient details of manpower, equipment and materials used will be recorded to enable a proper costing to be made at the conclusion of activities.

**11**. This plan will be reviewed annually and in the light of any e periencexgained or information received.

**REPORT ON OIL POLLUTION OBSERVATION**

When an oil slick on the water or oil pollution of the shore is observed it should be reported IMMEDIATELY to the County Oil Pollution Officer giving as many of the following details* as possible:

1. **Time and date of observation**
2. **Position of pollution** (give position relative to prominent or well-known place. If ashore give the nature of the beach)
3. **Extent of pollution** (area covered and 'thickness')
4. **Type of pollution** (if time allows, samples taken in bottles before treatment may pinpoint culprit)
5. **Source of pollution** (if known, name of ship installation, etc. If not known give apparent direction of origin).
6. **State of Tide and Sea** (flood or ebb tide. Rough, moderate or calm sea; any swell)
7. **Wind Direction and Force**
8. **Name and address of observer**
9. Names and phone numbers of County Oil Pollution Officers, office, home, deputies
10. **If the oil pollution is observed within the waters of a harbour there is a statutory obligation to report it to the Harbour Master**

| | |
|---|---|
| i.e. within Tralee Bay: | Cdr. Barrett, Fenit (066) 36103 |
| within Shannon Estuary: | Capt. E. K. Donnelly (061) 44914, home (061) 48862 |
| | Capt. J. O'Connor, Foynes 71 |
| | Capt. D. J. McLaughlin (061) 44914, home (061) 48474 |
| within Galway Bay: | Capt. J. Whyte (091) 2329, home (091) 2758 |

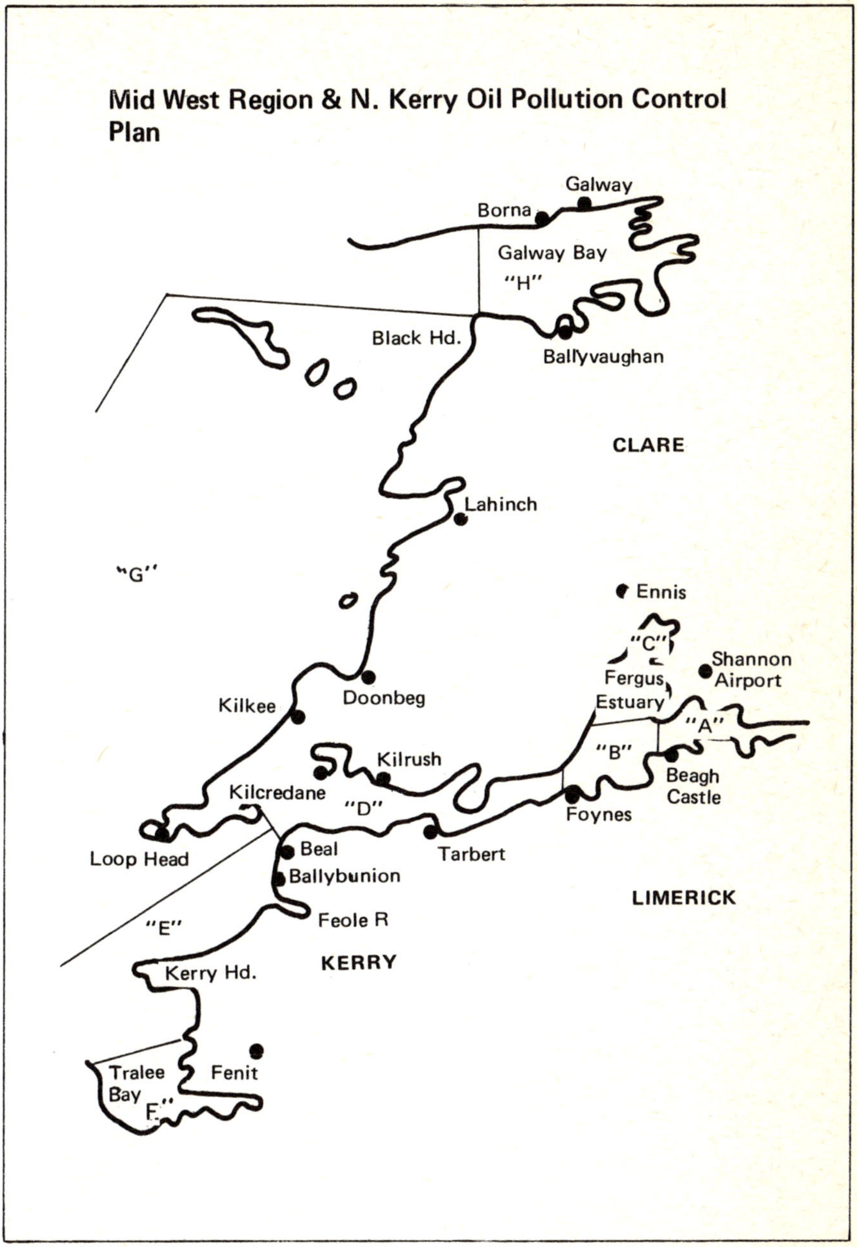
Mid West Region & N. Kerry Oil Pollution Control Plan
Galway
Borna
Galway Bay
"H"
Black Hd.
Ballyvaughan
CLARE
Lahinch
"G"
Ennis
"C"
Shannon Airport
Fergus Estuary
Kilkee
Doonbeg
"A"
"B"
Kilrush
Beagh Castle
Kilcredane
"D"
Foynes
Loop Head
Beal
Tarbert
Ballybunion
LIMERICK
Feole R
"E"
KERRY
Kerry Hd.
Tralee Bay
Fenit
"F"

| Ref. | Area | Physical features | Fish and crustsceans | Recommended action |
| --- | --- | --- | --- | --- |
| A. | Limerick to Beagh Castle | Fresh to salt shallow water and mud flats. Very strong currents. | Salmon, eels, young smelt etc. | Use no detergent. Physical removal where suitable and considered necessary*† |
| B. | Fergus Estuary | Salt water, varying depth, mud flats, very strong tides, especially between islands | Salmon, adult smelt, eels, etc. | ditto. |
| C. | Beagh Castle to Foynes | Salt water, mainly deep, mud flats, strong tides | ditto | As above except that low toxic detergents may be used freely during the first three hours of ebb tide. |
| D. | Foynes to Kilcredane/ Beal | Salt water, mainly deep. Stony beaches and mud flats. High amenity sandy beach at Beal. Strong tides. | Salmon, adult smelt in eastern portion. Sea fish and crustaceans found but not commercially fished. Lobster holding pond in Kilrush in summer months. | Use low toxic detergents from boats throughout. Early warning to be given in season to operator of Kilrush lobster pond of oil or detergent within 5 miles. Physical removal and/or treatment by dispersant from beaches. |
| E. | Beal Point to Brandon Head and off shore waters (excluding Tralee Bay) | Salt, mainly deep. Several high amenity sandy beaches. Strong tides. Subject to heavy swell and rough seas. | Salmon in Feale. Important crayfish area. Sea angling. Commercial sea fishing. | Physically remove and/or treat with dispersant any oil on bathing beaches. Oil slicks may be treated with detergent at sea but consultations with harbour and meteorological service officials important as tide and drift may take slick out of area |

| *Ref.* | *Area* | *Physical features* | *Fish and crustsceans* | *Recommended action* |
|---|---|---|---|---|
| F. | Tralee Bay | Salt moderate to shallow in depth. Extensive sandy bathing beaches and mud flats. | Most important oyster beds in Ireland. Also crayfish, lobster and commercial sea fishing | Strictly no detergents or sinking agents. Oil to be physically removed from beaches. Care to be taken not to disturb mud banks unduly |
| G. | Kilcredane to Loop Head to Black Head and offshore including Aran Islands | Salt, deep,mainly rocky foreshore – some inaccessible from land. Also some high amenity sandy beaches. | Lobster inshore crayfish offshore. Sea fishing. | Avoid sinking agents Detergents maybe used except within one mile of shore. Check possible drift with harbour and meteorological service. Bathing beaches may have oil physically removed and/or treated with detergent. |
| H. | Galway Bay | Salt, deep, mainly rocky foreshore with some important sandy bathing beaches, mud flats. | Lobsters, oysters, salmon and herring fishing in season. Commercial sea fishing. Lobster holding ponds at Barna and Ballyvaughan. | Strictly no detergents. Oil must be physically removed from beaches. |

* Cleaning of boats and gear with detergent is tolerable.

† On salt marshes oil patches should be left to nature

**The Directory of Pollution Control Equipment Companies in Western Europe**

compiled by R Whiteside

The second edition of this vital book will be expanded to cover the whole of western Europe

It will include details of more than 5000 companies and their equipment and services:

* address
* telephone and telex numbers
* key executives names
* description of products and services

The index breaks the companies down into 50 equipment and service categories.

Available Autumn 1976